APPLIED GYRODYNAMICS

WORKS OF

ERVIN S. FERRY

PUBLISHED BY

JOHN WILEY & SONS, Inc.

Applied Gyrodynamics.
A text-book for students and engineers. Develops the laws upon which gyroscopic devices are based, with many applications and much illustrative material, as well as a number of solved problems. xiv + 277 pages. 6 by 9. 218 figures. Cloth.

General Physics and its Application to Industry and Everyday Life.
A text-book for college students in general and technical courses in which especial emphasis is laid on the diverse relations of Physics to nature, agriculture, engineering and the home. Much of the motivation and illustrative material has not appeared heretofore in any text-book.
Third Edition, Revised. 839 pages. 5½ by 8. 603 figures. 1634 problems, many solved in detail. Cloth.

A Handbook of Physics Measurements.
By Erwin S. Ferry in Collaboration with O. W. Silvey, G. W. Sherman, Jr., D. C. Duncan, and R. B. Abbott. A self-contained manual of the theory and manipulation of those measurements in Physics which experience has shown to be most available for college and industrial laboratories.

    Vol. I.—Fundamental Measurements, Properties of Matter and Optics. Second Edition, Revised. xii + 288 pages. 5¼ by 8. 162 figures. Cloth.

    Vol. II.—Vibratory Motion, Sound, Heat, Electricity and Magnetism. Third Edition, Corrected. x + 277 pages. 5¼ by 8. 143 figures. Cloth.

Sensitive Element of Sperry Gyrocompass, Mark X (removed from binnacle).

APPLIED GYRODYNAMICS

FOR STUDENTS, ENGINEERS
AND USERS OF GYROSCOPIC APPARATUS

BY

ERVIN S. FERRY

Formerly Professor of Physics in Purdue University

NEW YORK
JOHN WILEY & SONS, Inc.
London: CHAPMAN & HALL, Limited

IN THE REPRINTING OF THIS BOOK, THE RECOMMENDATIONS OF THE WAR PRODUCTION BOARD HAVE BEEN OBSERVED FOR THE CONSERVATION OF PAPER AND OTHER IMPORTANT WAR MATERIALS. THE CONTENT REMAINS COMPLETE AND UNABRIDGED.

COPYRIGHT, 1932, 1933
BY ERVIN S. FERRY

All Rights Reserved

This book or any part thereof must not be reproduced in any form without the written permission of the publisher.

Printed in U. S. A.

PREFACE

Gyrodynamics is the brain child of the pure mathematician. From the beginning it was hedged about by a wall of differential equations. Some engineers and physicists have broken through this wall and have elicited the aid of gyrodynamics in producing marvels greater than those of fabled Daedalus. Already, through its aid, we have a device used on nearly all war ships and on many of the larger merchant ships by means of which the ship can be guided on any desired straight course without the aid of a helmsman; another that will keep an airplane level and on a straight course even though the aviator may be out of the cockpit making some necessary adjustment to the engine; another that will keep a submarine torpedo on either a straight or a curved course and, after the torpedo has run a predetermined distance, change the course by a predetermined angle; another that will prevent the rolling of a ship of any size even on a rough sea; another which, carried on a train running at ordinary speed, will record all irregularities of the track traversed. There are many important instruments used in aviation, navigation and in the industries that are based upon the principles of gyrodynamics. Stresses are developed in the structure of ships, airplanes and other vehicles carrying rotating machinery that can be computed and guarded against only by a designer familiar with gyrodynamics.

The purpose of the present book is to bring gyrodynamics out from behind the integral signs and to present it to the acquaintance of engineers and students having the mathematical equipment of the ordinary graduate of engineering or physics. This book is the outgrowth of lecture notes of a course that has been given for several years. The first stage in the development of the course was the collection of information of all gyroscopic devices of industrial importance. All of the United States patents and many foreign patents granted on gyroscopic apparatus since 1900 have been examined. Correspondence has been carried on with many firms and individuals. The test of a ship stabilizer in Philadelphia has been watched. Two factories making gyro-compasses in Brooklyn and the gyro-compass school in the Brooklyn Navy

Yard have been visited. Foreign-built gyro-compasses have been inspected on ships in New York harbor. One summer vacation of three months has been spent in attendance on the Sperry Gyro-compass School. A trip was made to Annapolis and Washington for the purpose of obtaining information on the gyro control of automobile torpidoes and a trip was made to Kiel, Germany, for the purpose of examining the construction of the Anschütz gyro-compass.

The second stage in the development of the course was the deduction of the laws and principles upon which depend the action of the various devices considered, using therein only methods that are understandable by students who are not specialists in mathematics. Equations are derived that can be used for the solution of numerical problems. Careful attention has been given to the construction of clear definitions and diagrams.

The first chapter of the present book is preliminary to the subject of gyrodynamics and includes the definitions and laws of physics assumed in the subsequent chapters. In the second chapter are developed the laws of gyrodynamics upon which are based the various gyroscopic devices used in industry. Many applications and much illustrative material are included as well as a number of solved problems designed to show methods of calculation. The gyroscopic pendulum with such applications as the gyro-horizon and the gyro-sextant are considered in the third chapter. The fourth chapter is devoted to the consideration of anti-roll devices for ships. In the fifth chapter the principles upon which the gyro-compass depends are developed, and each of the five makes of gyro-compass being installed on ships in 1931 is described in some detail. The last chapter considers the dynamic stabilization of a statically unstable body by means of a gyroscope. The methods which have been used for stabilizing monorail cars are described, and the reasons for the lack of success of the models built are indicated.

It is my agreeable duty to acknowledge the aid received from all to whom I have applied. For information concerning the gyro-compass I am indebted to Mr. E. C. Sparling, compass engineer, and to Mr. J. J. Brierly, head of the Gyro-compass School, of the Sperry Gyroscope Company, Brooklyn, N. Y.; to Mr. A. P. Davis, chief engineer of the Arma Engineering Company, Brooklyn, N. Y.; to M. P. Vonderwahl, chief engineer of Anschütz & Co., Kiel; to Mrs. S. G. Brown of the S. G. Brown Company, Limited, London,

England, and to the Director of the Officine Galileo, Florence, Italy. I am under deep obligation to Mr. F. P. Hodgkinson, stabilizer engineer, and to Mr. L. F. Carter, recorder engineer, of the Sperry Gyroscope Company, and to Mr. William Dieter, chief engineer of the E. W. Bliss Company, of Brooklyn, N. Y.

Professor Arthur Taber Jones of Smith College has been so good as to give the manuscript of the present book the same careful reading and scholarly criticism which he has bestowed on my previous books. These criticisms have resulted in a great improvement in the clearness and accuracy of the text.

ERVIN S. FERRY.

LAFAYETTE, INDIANA, U. S. A.
September 1, 1931.

CONTENTS

Preface.. iii

CHAPTER I

Definitions and Principles of Elementary Dynamics

§1. *Translation and Rotation*

ART.   PAGE

1. Linear Motion: Definitions and Units....................... 1
2. Angular Motion: Definitions and Units...................... 3
3. Composition and Resolution of Forces....................... 5
4. Composition and Resolution of Uniform Linear Velocities and Accelerations. Solved Problem........................... 7
5. Composition and Resolution of Torques..................... 10
6. Composition and Resolution of Angular Velocities, Accelerations and Moments... 10
7. The Relation between the Angular Velocity of a Body and the Linear Velocity of any Point of the Body................... 11
8. The Relation between the Angular Acceleration of a Body and the Linear Acceleration of any Point of the Body.............. 11
9. Centripetal and Centrifugal Forces.......................... 12
10. The Dynamic Vertical and the Dynamic Horizontal........... 12
11. Moment of Inertia.. 13
12. Values of the Moments of Inertia of Certain Bodies........... 15
13. Axes of the Principal Moments of Inertia of a Body........... 15
14. Centripetal Forces Acting on an Unsymmetrical Pendulum Bob... 16
15. The Relation between Torque and Angular Momentum........ 17
16. Conservation of Angular Momentum....................... 20
17. Centroid, Center of Gravity and Center of Mass.............. 21

§2. *Simple Harmonic Motion*

18. Simple Harmonic Motion of Translation..................... 22
19. The Period of a Simple Harmonic Motion of Translation....... 24
20. Simple Harmonic Motion of Rotation....................... 25
21. The Physical Pendulum................................... 28
22. The Conical Pendulum.................................... 29
23. Wave Motions and Wave Forms............................ 30
24. Phase and Phase Angle................................... 32
25. The Mean Value of the Product of Two Simple Harmonic Functions of Equal Period................................... 33
26. Oscillation of a Coupled System. Resonance................. 36
27. Damping of Vibrations................................... 39
28. The Frahm Anti-Roll Tanks............................... 42

CHAPTER II

The Motion of a Spinning Body Under the Action of a Torque

ART.		PAGE
29.	Degrees of Freedom...	44
30.	The Effect on the Motion of a Spinning Body Produced by an External Force...	45
31.	Precession...	48
32.	Change in the Motion of a Ship Produced by Precession of the Shaft	50
33.	Deviation of the Course of an Airplane Produced by Precessions of the Propeller Shaft.......................................	51
34–35.	Magnitude of the Torque Required to Maintain a Given Precessional Velocity when there is Zero Motion about the Torque-Axis, and when the Axes of Spin, of Torque and of Precession are Perpendicular to One Another. Solved Problems.........	52
36.	The Direction and Magnitude of the Torque Developed by a Spinning Body when Rotated about an Axis Perpendicular to the Spin-Axis. Solved Problems...............................	60
37.	The Period of Precession.....................................	66
38.	The Kinetic Energy of a Precessing Body......................	66
39.	Nutation...	67
40.	The Effect of Hurrying and of Retarding the Precession of a Spinning Body...	69
41.	Motion of the Spin-Axis Relative to the Earth.................	71
42.	The Angular Velocity of the Spin-Axle of an Unconstrained Gyroscope with Respect to the Earth. Solved Problem..........	72
43.	An Instrument for Measuring the Crookedness of a Well Casing..	74
44.	The Weston Centrifugal Drier.................................	75
45.	The Effect on the Direction of the Spin-Axis of a Top Produced by Friction at the Peg.......................................	76
46.	The Bonneau Airplane Inclinometer...........................	76
47.	The Sperry Airplane Horizon.................................	78
48.	The Drift of the Projectile from a Rifled Gun..................	80
49.	The Effect of Revolving a Non-Pendulous Gyroscope with Two Degrees of Rotational Freedom, about the Axis of the Suppressed Rotational Freedom...............................	82
50.	The Pioneer Turn Indicator...................................	85
51.	The Clinging of a Spinning Body to a Guide in Contact with It...	85
52.	Components of the Torque Acting upon a Spinning Body Having one Fixed Point, Relative to the Three Coördinate Axes of a Rotating Frame of Reference.............................	87
53.	Components of the Torque Required to Maintain Constant Precession of a Body when the Precession Axis is Inclined to the Spin-Axis..	90
54.	The Griffin Pulverizing Mill. Solved Problem.................	92
55.	The Automobile Torpedo.....................................	94
56.	The Pendulum-Controlled Depth Steering Gear................	95
57.	The Conditions that Must Be Fulfilled by the Horizontal Steering Mechanism...	96

CONTENTS

ART.		PAGE
58.	The Bliss-Leavitt Torpedo Steering Gear	97
59.	Method of Compensating the Effect of the Rotation of the Earth	99
60.	Devices for Changing the Course of a Torpedo	99
61.	Airplane Cartography	100
62.	Direct Control of the Axis of a Camera	101
63.	Indirect Control of the Axis of a Camera	101
64.	Control of the Line of Sight of a Camera	103

CHAPTER III
THE GYROSCOPIC PENDULUM OR PENDULOUS GYROSCOPE
§1. *General Properties*

65.	The Gyro-Pendulum	105
66.	The Period and the Equivalent Length of a Gyroscopic Conical Pendulum	105
67.	The Inclination of the Precession Axis of a Gyroscopic Conical Pendulum to the Vertical	106
68.	The Period of the Undamped Vibration, back and forth through the Meridian, of the Gyro-Axle of a Pendulous Gyroscope	108
69.	The Torque with Which the Second Frame of a Gyroscope Resists Angular Deflection	111
70.	The Length of the Simple Pendulum Which Has the Same Period as an Oscillating Body to Which is Attached a Spinning Gyroscope	113

§2. *Gyro-Horizontals and Gyro-Verticals*

71.	Determination of the Latitude of a Place	115
72.	Gyro-Horizons	116
73.	The Schuler Gyro-Horizon	117
74.	The Anschütz Gyro-Horizon	117
75.	The Bonneau-LePrieur-Derrien Gyro-Sextant	119
76.	Fleuriais Gyroscopic Octant	120
77.	The Sperry Roll and Pitch Recorder	121
78.	The Sperry Automatic Airplane Pilot	123
79.	The Sperry-Carter Track Recorder	125
80.	Directed Gun-Fire Control	128
81.	Gun-Fire Directorscopes	129

CHAPTER IV
GYROSCOPIC ANTI-ROLL DEVICES FOR SHIPS
§1. *The Oscillation of a Ship in a Seaway*

82.	The Rolling of a Ship due to Waves	131
83.	The Pitching of a Ship due to Waves	132
84.	The Metacentric Height	132
85.	The Experimental Determination of the Metacentric Height	133
86.	The Period of the Rolling Motion of a Ship	134
87.	Methods of Diminishing the Amplitude of Roll	135

CONTENTS

ART.   PAGE

§2. *The Inactive Type of Gyro Ship Stabilizer*

88. The Effect on the Motion of a Swinging Pendulum Produced by an Attached Gyroscope, (*a*) When the Precession of the Gyro-Axle is Opposed by a Frictional Torque................................ 136
89. The Effect of a Spinning Gyroscope on the Rolling of a Ship...... 140
90. The Schlick Ship Stabilizer..................................... 142
91. The Fieux Ship Stabilizer...................................... 144

§3. *The Active Type of Gyro Ship Stabilizer*

92. The Effect on the Motion of a Swinging Pendulum Produced by an Attached Gyroscope, (*b*) When the Gyro-Wheel is acted upon by a Torque about an Axis Perpendicular to the Spin-Axis and the Axis of Vibration of the Pendulum............................. 146
93. The Sperry Ship Stabilizer..................................... 147
94. Operation of the Sperry Ship Stabilizer......................... 149
95. The Braking System.. 151
96. Rolling of a Ship Produced by a Gyro........................... 152
97. Admiral Taylor's Formula. Solved Problem....................... 153

CHAPTER V
NAVIGATIONAL COMPASSES
§1. *The Various Types*

98. The Altitude and Azimuth Method of Locating the Geographic Meridian... 162
99. The Directive Tendency of a Magnetic Compass................... 165
100. The Deviations of a Magnetic Compass on an Iron Ship.......... 166
101. The Deviation of a Magnetic Compass Produced by a Rapid Turn 168
102. The Earth Inductor Compass.................................... 169
103. The Magneto Compass... 169
104. The Sun Compass... 170
105. The Apparent Motion of the Spin-Axle of an Unconstrained Gyroscope Due to the Rotation of the Earth................... 172
106. The Meridian-Seeking Tendency of a Pendulous Gyroscope........ 173
107. The Meridian-Seeking Tendency of a Liquid-Controlled Non-Pendulous Gyroscope.. 176
108. Making a Gyroscope into a Gyro-Compass........................ 177
109. The Meridian-Seeking Torque Acting on a Gyro-Compass.......... 178

§2. *The Natural Errors to Which the Gyro-Compass is Subject*

110. The Latitude Error.. 182
111. The Error Due to the Velocity of the Ship. The Meridian-Steaming Error... 182
112. The Deflection of the Axle of a Gyro-Compass Produced by Acceleration of the Ship's Velocity........................... 184
113. The Period of a Gyro-Compass that will have Zero Ballistic Deflection Error when at a Particular Latitude.................. 186
114. The Ballistic Damping Error................................... 190

CONTENTS

ART.		PAGE
115.	The Compass Error Due to Rolling of the Ship when on an Intercardinal Course. The Quadrantal or Rolling Error...........	191
116.	Quadrantal Deflection Due to Lack of Symmetry of the Sensitive Element..	195
117.	The Suppression of the Quadrantal Error.....................	195
118.	The Degree of Precision of Gyro-Compasses..................	196

§3. *The Sperry Gyro-Compass*

119.	The Principal Parts of the Master Compass...................	198
120.	The Follow-Up System	202
121.	The Method of Damping.....................................	203
122.	The Magnitude of the Latitude Error for which Correction Must Be made in the Sperry Gyro-Compass.......................	205
123.	Correction Mechanism for Velocity and Latitude Errors.........	206
124.	Avoidance of the Ballistic Deflection Error....................	210
125.	The Automatic Ballistic Damping Error Eliminator............	212
126.	Avoidance of the Quadrantal or Rolling Error.................	212
127.	The Repeater System..	213

§4. *The Brown Gyro-Compass*

128.	Production of the Meridian-Seeking Torque....................	215
129.	The Method of Damping.....................................	218
130.	Absence of Latitude Error...................................	220
131.	The Meridian-Steaming Error................................	220
132.	The Repeater System..	221
133.	The Ballistic Deflection Error................................	222
134.	Prevention of the Quadrantal or Rolling Error.................	222

§5. *The Anschütz Gyro-Compass*

135.	The Sensitive Element of the Model of 1926...................	224
136.	The Supporting System......................................	227
137.	Damping..	228
138.	The Meridian-Steaming Error................................	231
139.	Prevention of the Ballistic Deflection Error....................	231
140.	Avoidance of the Quadrantal or Rolling Error.................	232
141.	The Follow-Up Repeater System	234

§6. *The Arma Gyro-Compass*

142.	The Sensitive Element.......................................	236
143.	The Follow-Up System	238
144.	The Course and Speed Error Corrector........................	241
145.	Prevention of the Ballistic Deflection Error....................	244
146.	Avoidance of the Ballistic Damping Error.....................	246
147.	Avoidance of the Quadrantal or Rolling Error.................	247

§7. *The Florentia Gyro-Compass*

148.	Arrangement of the Principal Parts of the Master Compass......	247
149.	The Follow-Up System	248

ART.		PAGE
150.	Damping	250
151.	The Latitude and Meridian-Steaming Error Corrector	251
152.	Avoidance of the Ballistic Error	252
153.	Avoidance of the Error Due to Rolling and Pitching of the Ship When on Intercardinal Courses	252

CHAPTER VI
GYROSCOPIC STABILIZATION
§1. *General Principles*

154.	Static and Kinetic Stability	255
155.	The Stability of a System Consisting of a Body Capable of Oscillation and an Attached Precessing Gyro-Wheel	255
156.	Some Laws of Dynamic Stability	257

§2. *Gyroscopically Stabilized Monorail Cars*

157.	The Economy of Monorail Cars	259
158.	The Principles upon Which Depend the Dynamic Stabilization of Monorail Cars	259
159.	The Effect of a Change in Linear Velocity on the Stability of a Monorail Car that Carries a Single Statically Unstable Gyroscope with Vertical Spin-Axle	261
160.	The Effect of a Change in Linear Velocity on the Stability of a Monorail Car that Carries a Single Gyroscope with Horizontal Spin-Axle Transverse to the Car	264
161.	Methods for Increasing the Kinetic Stability of a Monorail Car While the Car is Going Around Curves	265
162.	The Schilovsky Monogyro Monorail Car	266
163.	The Brennan Duogyro Monorail Car	268
164.	The Scherl Duogyro Monorail Car of 1912	271

NOTATION USED

a Linear acceleration
a' Radial linear acceleration
$\mathbf{a}$ Angular acceleration
b Constant
c Constant
d_i Internal diameter
d_o Outside diameter
$\overline{d_1 d_2}$ Mean value of the product of d_1 and d_2
F Force
g Acceleration due to gravity
H Metacentric height
h_c Angular momentum with respect to an axis through c
h_s Angular momentum with respect to the spin-axis
K Moment of inertia
K_c Moment of inertia with respect to an axis through c
K_s Moment of inertia with respect to the spin-axis
k Radius of gyration
L Torque
l Length of the equivalent simple pendulum
λ Latitude
m Mass
P Power
P_H Horse-power
r Radius
R Resultant
S Torsional stiffness
t Time
T Period
v Linear velocity
v' Meridian component of velocity of ship
v_0 Linear velocity at time zero
v_t Linear velocity at time t
$_n v_s$ Linear velocity of a body s with respect to a body n
w Angular velocity
w_c Angular velocity about an axis through c
w_0 Angular velocity at time zero
w_t Angular velocity at time t
w_r Mean roll velocity of a ship
W Work or energy
W_t Kinetic energy of translation
W_{ro} Kinetic energy of rotation
x Linear distance or linear displacement

NOTATION USED

- β Angle — especially ballistic deflection
- δ_1 Meridian-steaming error
- ϕ, θ Angle or angular displacement
- Φ Mean angular displacement
- λ Latitude
- $\doteqdot$ Approximately equal to
- Σxy The sum of a series of terms of the form xy.

Different directed quantities are represented in diagrams by arrows having heads of different shapes. In the case of directed angular quantities, the arrow-head indicates that the torque, angular velocity, angular acceleration or angular momentum is in the clockwise direction about the length of the arrow as axis as seen by an observer looking along the shaft of the arrow toward the head. The arrow-heads used to represent the various directed quantities are as follows:

⟶	force (F)	angular momentum (h)	⟶
⟶	linear velocity (v)	torque (L)	⟶
⟶	angular velocity (w)	linear acceleration (a)	⟶

APPLIED GYRODYNAMICS

CHAPTER I

DEFINITIONS AND PRINCIPLES OF ELEMENTARY DYNAMICS

§1. *Translation and Rotation*

1. Linear Motion. Definitions and Units. — Linear displacements are commonly measured in centimeters, meters, kilometers, inches, feet, or miles of various lengths.

$$
\begin{aligned}
1 \text{ centimeter} &= 0.3937 \text{ inch} \\
1 \text{ meter} &= 3.2809 \text{ feet} \\
1 \text{ kilometer} &= 0.6214 \text{ statute mile} \\
1 \text{ statute mile} &= 5280 \text{ feet} \\
1 \text{ nautical mile} &= 6080 \text{ feet} \\
1 \text{ inch} &= 2.5399 \text{ centimeters} \\
1 \text{ foot} &= 0.3048 \text{ meter} \\
1 \text{ statute mile} &= 1.6093 \text{ kilometers}
\end{aligned}
$$

Linear velocity is time-rate of linear displacement in any given direction. The magnitude of velocity is called *speed*. Thus, one might say that an automobile is moving with a velocity of 30 miles per hour north but that the automobile is capable of developing a speed of 50 miles per hour. A speed of one nautical mile (6080 ft.) per hour is called a *knot*. If the linear displacement in every equal time interval t has the constant value x, then the magnitude of the constant linear velocity is

$$v = \frac{x}{t} \tag{1}$$

Linear acceleration is the time-rate of change of linear velocity. There is a linear acceleration whenever either the magnitude or the direction of a linear velocity changes. A body that in every time interval t changes in linear speed from v_0 to v_t while maintaining a constant direction of motion has a constant linear acceleration in the direction of motion of the value

$$a = \frac{v_t - v_0}{t} \qquad (2)$$

A body moving in a circular path of radius r with constant speed v has an acceleration directed toward the center of the circle. The magnitude of this so-called " radial acceleration " or " centripetal acceleration " is

$$a' = \frac{v^2}{r} \qquad (3)$$

Commonly employed units of linear acceleration are the centimeter per second per second, the foot per second per second, and the mile per hour per second.

Any cause which either produces or tends to produce a linear acceleration of the motion of a body is called *force*. Commonly employed units of force are the dyne, gram weight, and pound weight.

1 pound weight = 454 grams weight

At a place where the acceleration due to gravity is 980 centimeters per second per second

1 gram weight = 980 dynes

Work is the accomplishment of a change in the position of a body against an opposing force. The magnitude of the work is taken to be equal to the product of the force overcome and the projection of the displacement in the line of action of the force.

$$W = Fx \qquad (4)$$

The commonly employed units of work are the erg (dyne-centimeter), joule (10^7 ergs), and the foot-pound.

Energy is stored work. The amount of energy possessed by a system of bodies is the amount of work it can do in passing from its present condition to some standard condition.

The kinetic energy of translation of a body of mass m moving with a linear speed v is

$$W_t = \tfrac{1}{2}mv^2 \qquad (5)$$

Energy, being work, is measured in the same units as work.

Power is the time-rate of doing work. If a system does an amount of work W in time t by opposing a constant force F

through a distance x along the line of action of the force, the power P has the value

$$P = \frac{W}{t} = \frac{Fx}{t} = Fv \tag{6}$$

The units of power are the *erg per second*, the *watt* (joule per second), the *foot-pound per second*, and the *horse-power* (= 550 foot-pounds per second).

That property of a body which makes it necessary to use force to produce a linear acceleration in the motion of the body is called *inertia*. Inertia is measured by the tendency of a body to keep its linear velocity of constant magnitude and in an invariable direction. Anything which possesses inertia is called *matter*. The amount of matter in a body is called the *mass* of the body. The ratio of the masses of two bodies is taken to be equal to the ratio of the inertias of the bodies. The inertia of a body or system of bodies equals the sum of the masses of its parts. Commonly employed units of mass are the *gram*, the *pound*, and the *slug* or *British engineering unit of mass*. At a place where the acceleration due to gravity is 32.1 ft. per sec. per sec.,

$$1 \text{ slug} = 32.1 \text{ lb.}$$

The product of the mass of a body and its linear velocity is called the *linear momentum* of the body. The units used are the gram-centimeter per second and the slug-foot per second. These units have no name but are referred to as the centimeter-gram-second absolute unit and the British engineering unit of linear momentum, respectively.

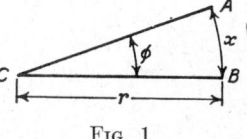

Fig. 1

2. Angular Motion: Definitions and Units. — Angular displacements are commonly measured in degrees, radians or revolutions. The *degree* is one-ninetieth part of a right angle. The *radian* is the plane angle subtended at the center of a circle by an arc equal to the radius of the circle. Thus, if the arc AB, Fig. 1, is half as long as the radius CA, the angle ϕ is one-half of a radian. Whatever the length x of the arc, and the length r of the radius

$$\phi = \frac{x}{r} \text{ radians} \tag{7}$$

$$1 \text{ revolution} = 2\pi \text{ radians} = 360°.$$
$$1 \text{ radian} \doteq 57.3° \doteq 3438'.$$

Angular velocity is the time-rate of change of angular displacement in a given sense about a given axis. The magnitude of angular velocity is called angular speed. Thus, one might say that a top has an angular velocity of 200 revolutions per minute in the clockwise direction about an axis inclined 30 degrees to the vertical and 45 degrees west of the meridian plane. If the angular displacement in every equal time interval t be constant and represented by ϕ, then the magnitude of the constant angular velocity is

$$w = \frac{\phi}{t} \qquad (8)$$

Angular acceleration is the time-rate of change of angular velocity. There is an angular acceleration whenever there is a change either in the magnitude of the angular velocity or in the direction of the axis of rotation. A body that in every equal time interval t changes in angular speed from w_0 to w_t while the direction of the axis of rotation remains unchanged, has a constant angular acceleration about the fixed axis of the value

$$\mathbf{a} = \frac{w_t - w_0}{t} \qquad (9)$$

Commonly employed units of angular acceleration are the radian per second per second and the revolution per minute per second.

Any cause which either produces or tends to produce an angular acceleration of the motion of a body is called a *torque* or *force couple*. A torque is equivalent to two equal, oppositely directed forces, acting in parallel lines. The magnitude of a torque is measured by the product of one of the forces and the perpendicular distance between the lines of action of the two forces. The line around which a torque either produces or tends to produce angular acceleration is called the *torque-axis*. The magnitude of a torque is expressed in centimeter-dynes, pound-feet, etc.

That property of a body because of which a torque is needed to give to the body an angular acceleration is called *moment of inertia*. Moment of inertia is measured by the tendency of a body to keep its angular velocity of constant magnitude and about an invariable axis of rotation. The moment of inertia of a particle of mass m at a distance r from the axis of rotation is mr^2. The moment of inertia of a body about a given axis is numerically equal to the sum of the products of the masses of the particles composing the body and the squares of their respective distances from the axis of rotation.

Units of moment of inertia are the gram-centimeter^2, the pound-foot^2, and the slug-foot^2. This last unit is the British engineering unit of moment of inertia.

When a body rotates against a constant opposing torque L through an angular displacement ϕ radians, it does an amount of work

$$W = L\phi \qquad (10)$$

The kinetic energy of a body rotating with angular velocity w is

$$W_{ro} = \tfrac{1}{2} Kw^2 \qquad (11)$$

where K represents the moment of inertia of the body with respect to the axis of rotation. If K is measured in feet and slugs, and w in radians per second, W_{ro} will be expressed in foot-pounds.

If an angular displacement ϕ be effected in time t with a constant angular velocity w, then the power developed by the torque is

$$P\left[= \frac{L\phi}{t} \right] = Lw \qquad (12)$$

The product of the moment of inertia of a body with respect to a given axis and the angular velocity of the body about the same axis is called the *angular momentum* of the body with respect to the given axis. Thus, in symbols, the angular momentum with respect to an axis through c is

$$h_c = K_c w_c \qquad (13)$$

It can be shown that the sum of the moments of the linear momenta of the elementary parts of a body equals the angular momentum of the body. For this reason, angular momentum is also called *moment of momentum*. It is sometimes called *kinetic moment*.

The units employed are the gram centimeter^2-radian per second and the slug foot^2-radian per second. They have no names but are referred to as the centimeter-gram-second absolute unit and the British engineering unit of angular momentum, respectively.

3. Composition and Resolution of Forces. — A force has both direction and magnitude. It can be completely represented by a straight line drawn in the direction in which the force acts, of a length proportional to the magnitude of the force, and having an arrow-head marked on the line pointing in the direction of the action of the force. Two quantities represented by lines that intersect are said to be *concurrent*. The motion of a body may be due to the effect of two or more simultaneous forces. Two systems of forces having identical effects on the motion of a body are said to be *equivalent*. A single force equivalent to two or more simultaneous forces is called the *resultant* of the set. Two forces which, acting together, are equivalent to a single force are called *components* of the given force. The operation of finding the resultant of a system of forces is called *composition* of forces.

The method of finding the resultant of two concurrent forces, called the parallelogram law, may be stated as follows: *If two adjacent sides of a parallelogram represent in direction and in magnitude two concurrent forces, both directed from the point of intersection, then the diagonal of the parallelogram drawn from this intersection represents completely, in direction and in magnitude, the resultant of the two forces.*

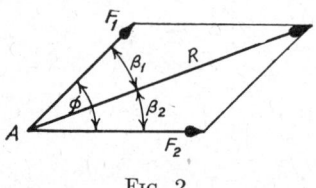

Fig. 2

If the magnitudes of the two concurrent forces be represented by F_1 and F_2, Fig. 2, and if the angle between them be ϕ, then from the law of cosines the magnitude of the resultant R is given by the equation

$$R^2 = F_1^2 + F_2^2 + 2 F_1 F_2 \cos \phi \tag{14}$$

The angle between the line representing either one of the component forces and the line representing the resultant can be obtained from Fig. 2 and the law of sines,

$$\frac{F_2}{R} = \frac{\sin \beta_1}{\sin (180° - \phi)} = \frac{\sin \beta_1}{\sin \phi}$$

$$\sin \beta_1 = \frac{F_2 \sin \phi}{R}$$

Similarly,

$$\sin \beta_2 = \frac{F_1 \sin \phi}{R}$$

$$\tag{15}$$

By compounding the resultant of two concurrent forces with a third force that is concurrent with this resultant, we can find the resultant of all three forces.

The operation of finding the components in two given directions of a force is called *resolution* of the force. The components of the force F, Fig. 3, in the directions A and B, Fig. 4, are F_A and F_B, Fig. 5; the components in the directions H and V, Fig. 4, are

F_H and F_V, Fig. 6. If H and V are at right angles to one another, $F_H = F \cos \phi$ and $F_V = F \sin \phi$.

A force may be resolved into three components in assigned directions. For instance, the components may be found in the vertical, in the north-south, and in the east-west directions.

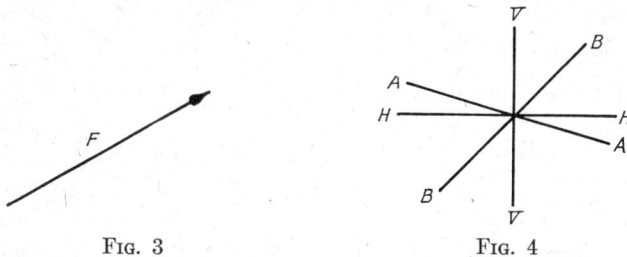

Fig. 3   Fig. 4

Linear velocities, linear accelerations, and linear momenta can be completely represented by directed straight lines as can forces. The parallelogram law, used for forces, can also be used to com-

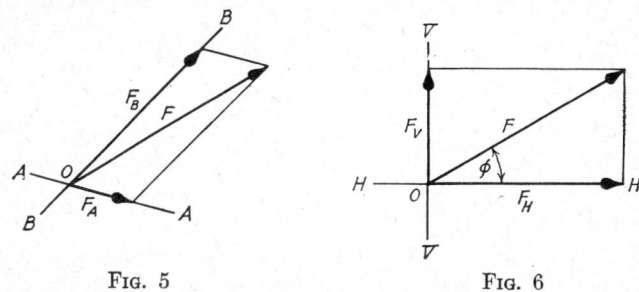

Fig. 5   Fig. 6

pound and resolve these quantities. However, since all velocity is relative to some reference body, care must be exercised to keep clearly in mind what object is moving and what object is taken as the body of reference. One method by which this can be done easily is illustrated in the following Article.

4. Composition and Resolution of Uniform Linear Velocities and Accelerations. — A passenger moving across a moving railway carriage furnishes an example of what are called simultaneous velocities. The passenger has a velocity relative to a point of the carriage, and at the same time the carriage has a velocity with respect to a point on the earth. These are called components of the passenger's motion. The velocity of the passenger with reference to the earth is called his resultant velocity.

In order that we may keep clearly in mind what object is re-

8　PRINCIPLES OF ELEMENTARY DYNAMICS

garded as moving and what object is taken as the reference body, it is convenient to use two subscripts. Let a subscript written at the right indicate the moving body, and a subscript written at the left indicate the frame of reference. Thus, the velocity of a tor-

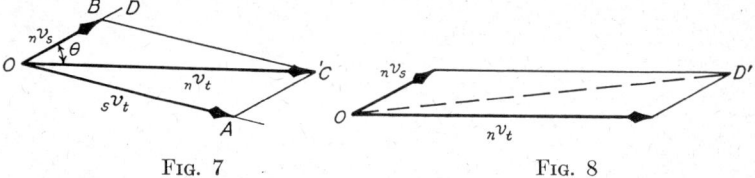

Fig. 7   Fig. 8

pedo with reference to the ship from which it is projected may be written $_sv_t$, and the velocity of the ship with reference to the enemy may be written $_nv_s$. In Fig. 7, OA represents $_sv_t$ and OB represents $_nv_s$. We notice that in these symbols for the velocities which we have compounded, the right subscript of one velocity is the same as the left subscript of the other, and that the resultant has the two subscripts that are different. In our example, $_sv_t$ and $_nv_s$ combined give the velocity of the torpedo with respect to the enemy, $_nv_t$.

If we have given $_nv_t$ and $_nv_s$ and wish $_sv_t$ we need only notice that $_nv_s = -_sv_n$. Thus, we can compound $_nv_t$ and $-_sv_n$ thereby obtaining $_sv_t$.

The "torpedo director" consists of two rods variable in length and direction which are linked together and to a third rod carrying a telescope. The telescope points in the direction $_nv_t$ when the other two rods are adjusted to represent in direction and length $_sv_t$ and $_nv_s$. When the enemy ship arrives at C, Fig. 7, it is seen in the telescope fastened to the rod OC. At this instant the torpedo is launched in the direction and with the speed relative to the firing ship represented by OA. The torpedo meets the enemy ship at A.

If, however, we attempt to combine in this manner two velocities which do not have a right subscript of one the same as a left subscript of the other, we shall have a result which is without meaning. For example, if we attempt to combine the velocity of the torpedo with respect to the enemy and the velocity of the firing ship with respect to the enemy, we would get the line OD', Fig. 8. This line has no physical meaning.

Problem. Waves are moving across the sea with a velocity of one knot westward, relative to the earth. The distance between the crests is 50 ft.

TRANSLATION AND ROTATION

Find the number of wave crests that in one hour pass under a ship which is steaming at a velocity of 15 knots: (a) north-east; (b) south-west.

Ans. (a) 1909 crests; (b) 1739 crests.

Problem. An airplane has a supply of gasoline sufficient for six hours of flying at a speed of 60 knots relative to the air. The wind has a velocity of 15 knots from the south. The pilot is ordered to proceed as far as possible along a north-east course and then retrace his course to the starting point. Find: (a) the course and ground speed to the turning point; (b) the course and ground speed back; (c) the distance from the starting point to the turning point; (d) the time from the starting point to the turn.

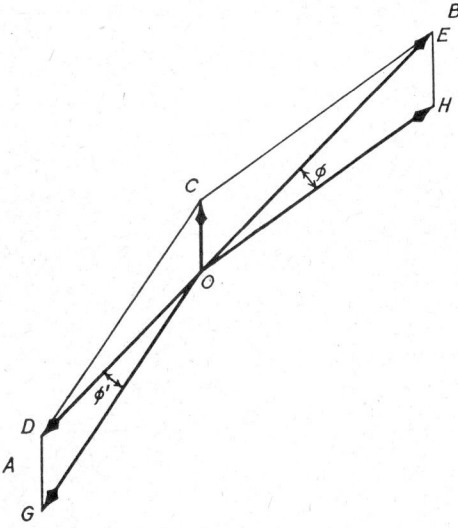

Fig. 9

Outline of Solution. Draw OC, Fig. 9, representing the velocity of the wind relative to the earth ($_ev_w$ = 15 mi. per hr.). From C describe two arcs of a circle of a radius ($CD = CE$) representing the speed of the plane relative to the wind ($_wv_p$ = 60 knots). Draw lines CD and CE to the points at which these arcs intersect the ground course AB. Complete the parallelogram CG and CH. The course out is along the line OH and that back is along OG. The ground velocity ($_ev_p$) is given by OE and the ground velocity back ($_ev_p'$) is represented by OD.

(a and b) The data of the problem give the magnitude $OH = OG$ together with the angles COB and COD. Using the law of sines and the law of cosines we find the values of ϕ, ϕ', OE which represents the ground velocity out ($_ev_p$), and OD which represents the ground velocity back ($_ev_p'$).

(c) We have two expressions for the distance to the turning point:

$$x = {_ev_p}t$$
$$x = {_ev_p}'t'$$

where t represents the time occupied in going to the turning point and t' represents the time occupied in returning. Then, the time of the round trip

$$t + t' = \frac{x}{e^v p} + \frac{x}{e^v p'} = \frac{x(e^v p + e^v p')}{e^v p \, e^v p'}$$

Whence, the ground distance to the turn

$$x = \frac{(e^v p \, e^v p')(t + t')}{e^v p + e^v p'}$$

(d) The time from the starting point to the turn is given by the expression

$$t = \frac{x}{e^v p}$$

5. Composition and Resolution of Torques. — Three quantities are required to specify completely a torque. They are the magnitude of the torque, the direction of the torque-axis, and the sense in which the torque acts around that axis, that is, whether it is clockwise or counter-clockwise. A torque can be completely represented by a straight line of a length proportional to the magnitude of the torque, drawn in the direction of the torque-axis, and carrying an arrow-head so placed that on looking along the line in the direction of the arrow, the sense of the torque is clockwise.

The method of compounding concurrent torques is as follows: *Draw from a given point two lines that represent the two simultaneous instantaneous torques. Complete a parallelogram on these two lines as sides. Then, the diagonal of the parallelogram drawn from the given point represents completely the resultant instantaneous torque.*

6. Composition and Resolution of Angular Velocities, Accelerations and Momenta. — An angular velocity, an angular acceleration, or an angular momentum can be represented completely by a straight line of a length proportional to the magnitude, drawn in the direction of the axis and carrying an arrow-head which indicates the sense of the angular quantity. In this book the arrow-head is so placed that, on looking along the line in the direction of the arrow, the sense of the angular velocity, acceleration or momentum is clockwise.

Angular velocities, angular accelerations and angular momenta are compounded and resolved by the parallelogram law exactly as are linear velocities and accelerations.

Experiment. — Set the Maxwell Top, shown in section in Fig. 10, into spinning motion when the point of support is (a) below the center of gravity of the top, (b) above the center of gravity. Observe that the spin-axis of the body rotates in the direction of the spin when the center of gravity is above the point of support, and in the direction opposite the spin when the center of gravity is below the point of support. For each case make an angular momentum diagram showing the angular momentum at the beginning and end of a brief time interval and the change of angular momentum during this interval.

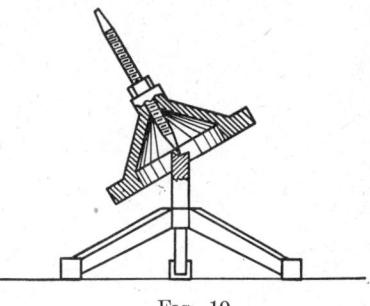

Fig. 10

7. The Relation between the Angular Velocity of a Body and the Linear Velocity of Any Point of the Body. — Consider a body, Fig. 11, rotating with uniform angular velocity w about an axis through O perpendicular to the plane of the paper. Let XX' be a line fixed in space and PO a line fixed in the body perpendicular to the axis of rotation. If the body rotates steadily through the angle ϕ in t seconds, then the angular speed is given, (1), by

$$w = \frac{\phi}{t} = \frac{x}{rt} = \frac{v}{r} \text{ radians per sec.} \qquad (16)$$

where x represents the arc of radius r subtended by the angle ϕ, and v represents the linear speed of the point P. This equation shows that the angular speed of a body equals the linear velocity of any point of the body divided by the distance of that point from the axis of rotation.

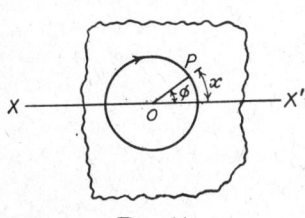

Fig. 11

8. The Relation between the Angular Acceleration of a Body and the Linear Acceleration of Any Point of the Body. — If the body be acted upon by a torque, the angular velocity about the torque axis will change in t seconds from a value w_o to some value w_t. The rate of change of angular velocity, or the angular acceleration a, is

$$\mathbf{a}\left[= \frac{w_t - w_o}{t} \right] = \frac{\frac{v_t}{r} - \frac{v_o}{r}}{t} = \frac{v_t - v_o}{rt} = \frac{a}{r} \qquad (17)$$

where a represents the tangential component of the linear acceleration of a point distant r from the axis of rotation. Thus, in words, the tangential component of the angular acceleration of a body equals the linear acceleration of any point of the body divided by the distance of that point from the axis of the acceleration.

9. Centripetal and Centrifugal Force. — It is readily shown* that, when a body is acted upon by a force of constant magnitude having a line of action always in the same plane and always perpendicular to the direction of the motion, (1) the speed of the body does not change, (2) the path of the body is a circle, (3) the body is moving with a linear acceleration which is always directed toward the center of the circle, and (4) the magnitude of this radial acceleration is constant and equal to the square of the linear speed of the body divided by the radius of its path.

According to Newton's First Law of Motion, a moving body will continue to move in a straight line with constant speed until acted upon by an external force. From the preceding paragraph, when a body of mass m moves with constant speed v in the circumference of a circle of radius r, we see: (1) that there must be a force acting upon the body, (2) this force must be directed toward the center of the circle, and (3) the magnitude of the force is given by

$$F_c \, [= ma] = m \frac{v^2}{r} \qquad (18)$$

The force required to overcome the inertia of a body in deflecting it from a rectilinear path into a circular path is called *centripetal* force. The agent which constrains the body to move in a circular path is acted upon by a force directed away from the center of the circle and has a magnitude equal to the force acting upon the body that is moving in the circular path. This, frequently called *centrifugal* force, is the reaction of the centripetal force and may be defined as the resistance which the inertia of a body in motion opposes to whatever deflects it from the rectilinear path. The centripetal force acts upon the body that is moving in a circular path, and its reaction, the centrifugal force, acts, not upon this body, but upon the agent which constrains it to move in a circular path.

10. The Dynamic Vertical and the Dynamic Horizontal. — A plumb-bob that is at rest or in uniform linear motion relative to the earth indicates the direction of the force of gravity at the

---

* Ferry's General Physics, Art. 66.

place where it is situated. This direction is the *true vertical*. When a plumb-bob is moving with a linear acceleration relative to the earth, it assumes a direction called the *dynamic vertical* corresponding to the given linear acceleration.

A plane normal to the true vertical is called a *true horizontal*. A plane normal to a dynamic vertical is called a *dynamic horizontal* corresponding to the given linear acceleration. The free surface of an unaccelerated liquid is truly horizontal. The free surface of a liquid in a vessel moving with linear acceleration is not truly horizontal. When the speed of a locomotive is increasing, the water gauge gives an indication that is too high; when the speed is decreasing, the indication is too low. The free surface of water in a rotating vessel is not truly horizontal.

11. Moment of Inertia. *Experiment.* — With the arm outstretched, rotate back and forth an iron pipe on which are mounted two iron balls, Fig. 12. Note that a considerable torque, produced by twisting the wrist, must be given the rod to change quickly the angular velocity, whereas a smaller torque is required to produce a smaller angular acceleration. Note that when the spheres are close to the hand, a fairly small torque, produced by twisting the wrist, will impart to the apparatus a certain angular acceleration about the outstretched arm as an axis. When the spheres are at a distance of a foot or more from the hand, a much larger torque is required to produce the same angular acceleration.

Fig. 12

That property of a body because of which a torque is needed to give the body an angular acceleration is called *moment of inertia*. When a torque acts upon a body, it produces an angular acceleration that is directly proportional to the torque and inversely proportional to the moment of inertia of the body upon which it acts. The moment of inertia is to be taken about the axis of the torque. The angular acceleration is about the same axis. Or, in symbols,
$$L = K\mathbf{a} \tag{19}$$

The moment of inertia K of a body consisting of particles of masses m_1, m_2, m_3, etc., at distances r_1, r_2, r_3, etc., respectively, from the axis of rotation is
$$K = m_1 r_1^2 + m_2 r_3^2 + \text{etc.}$$
or, in briefer notation
$$K = \Sigma(mr^2). \tag{20}$$

If mass be expressed in slugs and distance in feet, the moment of inertia will be expressed in British engineering units sometimes called slug-feet².

The moment of inertia of a body with respect to one axis has not the same value as the moment of inertia of the same body with respect to a different axis. The moment of inertia of a given body with respect to a given axis is numerically equal to the mass of a body which, if concentrated at unit distance from the axis of rotation, would require the same torque as the original body to produce the same angular acceleration. For any rigid body revolving about any axis fixed in space, the moment of inertia with respect to that axis is a constant quantity quite independent of both the speed of rotation and the torque acting.

The distance from the axis of rotation at which the entire mass of a body might be concentrated without altering the moment of inertia of the body with respect to that axis, is called the *radius of gyration*, or *swing radius*, of the body about the given axis. If the entire mass M of the body were at the distance k from the axis, the moment of inertia of the body with respect to this axis would be

$$K = Mk^2 \qquad (21)$$

In this equation, k is the radius of gyration of the body.

Experiment. — Clamp the frame of a bicycle in an upright position with the front wheel off the ground. With the wheel not spinning, apply a torque to the handle-bars so as to impart to the front wheel an angular acceleration about the handle-bar post as an axis. Now set the front wheel spinning and apply a torque to the handle-bars as before. Note that the torque to produce a given angular acceleration is much greater when the wheel is spinning than when it is not.

The torque required to impart to a spinning body a given angular acceleration about an axis perpendicular to the spin-axis is greater than the torque required when the body is not spinning. The ratio of the torque applied to a spinning body about an axis perpendicular to the spin-axis, to the angular acceleration thereby produced about the torque axis, is sometimes called the *dynamic moment of inertia* of the spinning body with respect to the torque-axis. The magnitude of the dynamic moment of inertia of a body with respect to the torque-axis depends upon the angular momentum of the part of the body that is spinning, with respect to the spin-axis.

12. Values of the Moment of Inertia of Certain Bodies. — The moments of inertia of a body, having regular geometrical shape, can be computed, but the moments of inertia of a body of irregular shape are usually most easily determined by experiment. The experimental determination is usually made by comparison with a body of known moment of inertia. For such comparisons cylinders and rings of known dimensions are convenient.

It is shown in books on Mechanics that the moment of inertia of a uniform right solid cylinder of mass m and diameter d, about its geometric axis, is

$$\tfrac{1}{8} md^2 \qquad (22)$$

whereas about any axis parallel to the geometric axis and distant p from it, the moment of inertia is

$$\tfrac{1}{8} md^2 + mp^2 \qquad (23)$$

If the cylinder has a length x, the moment of inertia about an axis through the center and normal to the length is

$$m\left[\frac{d^2}{16} + \frac{x^2}{12}\right] \qquad (24)$$

whereas about an axis coinciding with the diameter of one end, the moment of inertia of the cylinder is

$$m\left[\frac{d^2}{16} + \frac{x^2}{3}\right] \qquad (25)$$

The moment of inertia of a ring or right hollow circular cylinder of outer diameter d_o and inner diameter d_i, with respect to the geometric axis, is

$$\tfrac{1}{8} m[d_o^2 + d_i^2] \qquad (26)$$

If the moment of inertia of a body of mass m about an axis through the center of mass of the body is K_c, then the moment of inertia about any parallel axis distant p from the first axis is

$$K_p = K_c + mp^2 \qquad (27)$$

13. Axes of the Principal Moments of Inertia of a Body. — The moment of inertia of a sphere about an axis through the center of mass equals the moment of inertia about any other axis through the same point. The moment of inertia of a right cylinder about an axis through the center of mass and perpendicular to the length of the cylinder is greater than the moment of inertia about any other axis through the same point if the length of the cylinder is greater than the diameter. The moment of inertia of the cylinder about an axis through the center of mass and in the direction of the length is less than that about any other axis. In the case of most bodies, the values of the moments of inertia relative to their mutually perpendicular axes are unequal.

It can be shown that the axis of greatest and that of least moment of inertia are perpendicular to one another. The perpendicular axis through the center of mass of a body about which the moment of inertia is maximum, that about which it is minimum, and the other axis perpendicular to these two, are called the *principal momental axes* of the body. The moments of inertia about the principal momental axes are called the *principal moments of inertia* of the body. If the principal moments of inertia of one body or system are equal, respectively, to the principal moments of inertia of another body or system, the two bodies or systems are said to be *equimomental*.

For any rigid body there can be constructed an equimomental system consisting of three slender and uniform rigid rods bisecting each other at right angles, and coinciding in direction with the principal momental axes of the given body.

14. Centripetal Forces Acting upon an Unsymmetrical Pendulum Bob. — Consider a pendulum having a bob that is capable of rotation with negligible friction about the axis of the pendulum rod and that is unsymmetrical with respect to the axis of the pendulum rod. In Fig. 13, the pendulum bob is a rectangular bar with the long axis perpendicular to the pendulum rod. The long axis of the bob is inclined at an angle ϕ to the plane of the knife-edge and the pendulum rod. Consider A and A' at the centers of the two ends of the bob at the instant when the pendulum is passing through the equilibrium position. From A and A' draw lines AB and $A'B'$ perpendicular to the line CC' through the center of the bob parallel to the knife-edge. From B and B' draw lines BD and $B'D'$ perpendicular to the knife-edge. The points D and D' are the points about which oscillate the two particles at A and A'.

In order that the particles at A and A' may rotate about DD', the particles must be acted upon by centripetal forces F_c and F_c' directed toward D and D', respectively. Each of the centripetal forces F_c, etc., can be resolved into three components, one vertical, one parallel to the axis of the bob, and one horizontal and perpendicular to that axis. In the diagram, the horizontal components acting on the particles A and A', perpendicular to the axis of the rod, are marked $_hF_c$ and $_hF_c'$, respectively. The horizontal force acting on each particle necessary to cause the bob to swing so that the long axis is inclined at an angle ϕ to the plane through the knife-edge and pendulum rod is non-existent when there is zero friction between the bob and the pendulum rod. If these horizontal forces are non-existent, the inertia of each particle

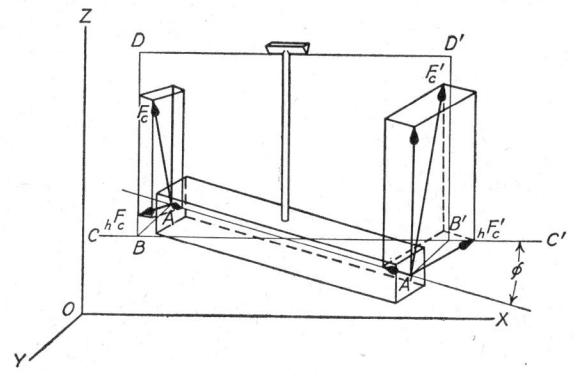

Fig. 13

of the bob will cause it to retreat as far as possible from the line CC'. Therefore the stable position of the loose bob is attained when the long axis of the bob is perpendicular to the plane of the knife-edge and pendulum rod.

An oscillating body tends to set itself in the position in which its moment of inertia with respect to the vibration axis is maximum. A body having equal principal moments of inertia will oscillate in any plane through the center of mass without any tendency to turn.

Similarly, if a body, supported at the center of mass so as to be capable of turning in any direction, be moved in a curved path, the body will tend to turn so that the axis of minimum moment of inertia is tangent to the path. A body that has equal principal moments of inertia has no tendency to turn when moving in a curved path.

15. The Relation between Torque and Angular Momentum. — The product of the moment of inertia of a body K_c with respect to an axis through a point c, and the angular velocity w_c about the same axis, is called the *angular momentum* of the body with respect

18 PRINCIPLES OF ELEMENTARY DYNAMICS

to the same axis. Thus, in symbols, the angular momentum with respect to an axis through a point c

$$h_c = K_c w_c$$

If a rigid body rotates with an angular acceleration **a**, about an axis fixed in space, it must be acted upon by a torque (19):

$$L = K\mathbf{a}$$

where K is the moment of inertia of the body with respect to the axis of torque. Since angular acceleration is the rate of change of angular velocity, we may write,

$$L\,[= K\mathbf{a}] = \frac{K\,dw}{dt} \tag{28}$$

The numerator of the right-hand member of this equation is the change of angular momentum. The right-hand member is the time-rate of change of angular momentum. The above equation shows that:

(a) The resultant torque acting on a rigid body about an axis fixed in space, equals the time-rate of change of the angular momentum of the body relative to the given axis.

(b) There can be no increase or diminution of the angular momentum of a body about an axis in space without the action of a corresponding external torque about that axis. This theorem is the analogue in rotation to Newton's First Law of Motion.

(c) When a torque acts upon a body, it produces an angular acceleration, and a rate of change of the angular momentum of the body, about the axis of torque, proportional to the torque.

(d) When a torque acts upon a body, there acts upon some other body another torque of the same magnitude in the opposite direction about the same axis. This theorem is the analogue in rotation of Newton's Third Law of Motion.

(e) When the angular momentum of a body is constant about any axis, then the resultant torque about that axis is zero.

Suppose that at a certain instant a body is spinning with constant angular speed about an axis in the direction of the line h, Fig. 14. The arrow-head on this line indicates the direction of spin about the spin-axis, and the length of the line is proportional to the product of the angular speed and the moment of inertia of the body about the spin-axis. This line represents completely the angular momentum of the body about the spin-axis at the

TRANSLATION AND ROTATION

chosen instant. Suppose that during a short time interval Δt afterward the direction of the spin-axis changes by a small angle $\Delta \phi$ and the magnitude of the angular momentum remains constant. In Fig. 14, h' represents the angular momentum at the end of this brief time interval, and Δh represents the change of angular momentum during this interval. The rate of change of angular momentum about the spin-axis is $\frac{\Delta h}{\Delta t}$. This is the value of the

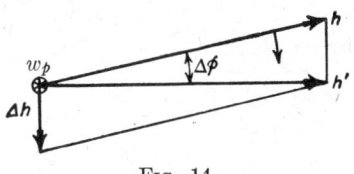

Fig. 14

torque about the axis Δh that is required to change the direction of the angular momentum from h to h'. This torque produces a rotation w_p about an axis perpendicular to the plane of the torque-axis and the spin-axis. Thus, *a torque acting upon a spinning body about an axis perpendicular to the spin-axis causes the spin-axis to rotate about another axis perpendicular to the plane of the axes of spin and of torque.* This is analogous to the case, in translation, in which a force of constant magnitude, acting perpendicular to the direction of motion of a body of constant linear speed, causes the body to describe a circular path.

Experiment. Stand on a rotatable table and hold above the head a bicycle wheel with axle vertical. The wheel should be provided with a rim consisting of 20 to 25 pounds of iron wire. Set the wheel into rotation by pressing against the spokes with a pointed stick. Note that while applying a torque to the wheel, the rotatable table turns in the opposite direction.

Consider a rod suspended at the center of mass by a string. Let the axis of the rod make an angle θ to the vertical. Suppose that a blow be imparted to one end of the rod in a direction perpendicular to the plane of the string and the axis of the rod. At any instant there is an angular momentum h about an axis perpendicular to the rod and in the plane containing the string and the axis of the rod. The angle θ is changing and h is changing both in magnitude and in direction. If a force couple of the proper constant magnitude be applied about an axis which is always perpendicular to the plane of h and the axis of the rod, then θ will remain of constant magnitude. The rod now is rotating with constant angular velocity about the string as axis, tracing the surfaces of two cones having a common apex at the center of gravity of the rod. This applied couple constitutes a centripetal torque. It is producing angular momentum about its own axis perpendicular to the plane containing h and the axis of the rod. This angular momentum, combining with the h of the previous position of the rod, gives the h in the new position. We have here an example of the case in which the axis of the resultant angular momentum does not coincide with the axis of the resultant angular velocity.

16. Conservation of Angular Momentum. — In the special case in which a rigid body is acted upon by a system of torques the sum of which, relative to a given axis, is zero, we have (28):

$$L = \frac{K\,dw}{dt} = 0$$

That is, in this special case, the time-rate of change of angular momentum is zero. In other words, the angular momentum is constant. Therefore, if a rigid body is acted upon by a set of external torques, the resultant of which about any assigned axis fixed in space is zero, the angular momentum of the body relative to the assigned axis is a constant quantity. From this equation it follows that:

However the parts of any material system may act upon one another, the total angular momentum of the system, about any axis fixed in space, will remain constant, so long as the system is acted upon by no outside torque. This theorem is called the Principle of the Conservation of Angular Momentum.

If a circus performer wishes to turn a series of aerial somersaults, he will jump from a springboard in such a way that he will give himself a considerable angular velocity about a horizontal axis. While turning the first somersault he will hold his body nearly straight, but after that he will gradually draw his head and knees together, thereby decreasing the moment of inertia of his body about the axis of rotation. Since the angular momentum of his body remains constant, when the moment of inertia is diminished the angular velocity of his body is increased to such an extent that he can make more somersaults before reaching the ground. If, when he has nearly reached the ground, he sees that he will fall face downward, he will straighten his body, thereby increasing his moment of inertia, thus decreasing his angular velocity to such an extent that he can alight on his feet.

Experiment. — Set into rotation about a vertical axis the horizontal rods carrying the two metal spheres, Fig. 15. Pull on the handle *H*, thereby drawing the spheres close to the axis of rotation and diminishing the moment of inertia of the rotating system. Note that the angular velocity of the system is now much greater than before. Release the pull on the handle. The spheres retreat from the axis of rotation, thereby increasing the moment of inertia of the rotating system. Note that the angular velocity diminishes. During these operations no torque was applied to the rotating body about the axis of rotation. Consequently, there was no change in the angular momentum of the system.

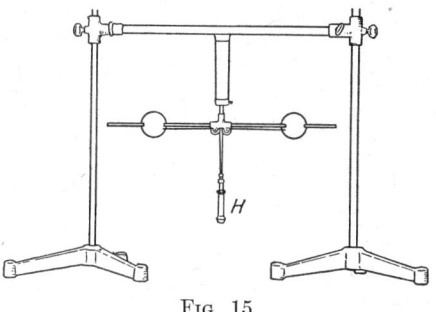

Fig. 15

Although the work done in pulling the spheres toward the axis of rotation does not change the angular momentum (Kw) of the rotating system, it does increase the kinetic energy of rotation ($\frac{1}{2} Kw^2$) of the system. While approaching the axis of rotation, the spheres are moving not in a circular path but in a spiral path. The force in the string is not perpendicular to the path. There is a component of this force that acts upon the spheres in the direction of the tangent. This tangential component of the force produces an increase in the tangential velocity and also in the angular velocity.

17. Centroid, Center of Gravity and Center of Mass. — The point of application of the resultant of a system of parallel forces is called the *centroid* or center of the system of forces.

"If the action of terrestrial or other gravity on a rigid body is reducible to a single force in a line passing always through one point fixed relatively to the body, whatever be its position relative to the earth or other attracting mass, that point is called the *center of gravity*."* The only bodies which have true centers of gravity are uniform spherical shells, uniform spheres, and spheres whose density changes from the center to the circumference according to some definite law. In the case of other bodies, the line of action of the weight of the body does not pass through the same point when the position of the body is changed. In engineering, however, it is customary to assume that there is a definite point

---

* Thompson and Tait, Natural Philosophy, II, p. 78.

fixed in the body at which the weight of the body is applied. Thus, it is assumed that every body has a center of gravity coincident with the point at which the resultant of the gravitational forces acting on the body would be applied if they were parallel.

If a body or system of bodies be conceived to be divided into particles of equal mass, then that point, the distance of which from any given plane is equal to the average distance from that plane of all the constituent particles, is termed the *center of mass*, or *center of inertia*, of the body or system of bodies.

If we represent by m_1, m_2, etc., the masses of the particles composing a body of total mass M, and by x_1, x_2, etc., the respective distances of these particles from any given plane, then it can be shown that the distance of the center of mass of the body from the given plane is

$$X = \frac{m_1 x_1 + m_2 x_2 + \text{etc.}}{M} \qquad (29)$$

The center of mass has the following properties:

(*a*) The center of mass is coincident with the centroid.

(*b*) The weight of a body acts approximately at its center of mass.

(*c*) If the resultant of all the forces acting on a rigid body is a single force the line of action of which passes through the center of mass of the body, the motion of the body is without angular acceleration.

(*d*) The center of mass of a body is so situated that the linear motion of the body will not be changed if the total mass were concentrated at this point and all the forces acting on the body were transferred to this point without change of magnitude or direction.

(*e*) The motion of the center of mass of any material system is not affected by the internal forces between the parts of the system, but only by external forces.

(*f*) No material system can of itself, without the action of external forces, change the motion of its center of mass.

(*g*) An unconstrained body, when acted upon by a system of forces equivalent to a couple, will rotate about its center of mass with constant angular acceleration.

§2. *Simple Harmonic Motion*

18. Simple Harmonic Motion of Translation. — When a body moves with a reciprocating motion which has at every instant a linear acceleration directed toward the center of the path and that

SIMPLE HARMONIC MOTION

varies directly with the distance of the moving body from that point, the body is said to have *a simple harmonic motion of translation*.

The defining equation is

$$a = -cd \qquad (30)$$

where c is a positive constant. The negative sign indicates that displacement d is measured from the equilibrium position whereas acceleration a is directed toward the equilibrium position.

Any elastic body that is distorted within the limit for which Hooke's Law is true, and then released, will thereafter vibrate with simple harmonic motion of translation. The motions of many other bodies are resultants of simple harmonic motions.

The time which elapses between two consecutive passages of the oscillating body in the same direction through any given point of its path is called the *period* of the motion. The maximum distance attained by the oscillating body from its position of equilibrium is called the *amplitude* of the motion.

It will now be shown that if a point moves with uniform speed in the circumference of a circle, the projection of the point on any straight line in the plane of the circle moves with simple harmonic motion of translation. This fact is the basis of our most easy methods for finding values of the period of a simple harmonic motion of translation as well as values for the displacement, acceleration and velocity of a body vibrating with simple harmonic motion.

Let a particle P', Fig. 16, move with uniform speed in the circumference of a circle $P'A'B'$, of radius r, and let P be the projection of this point on any right line AB in the plane of the circle. As P' moves with uniform speed in the circumference of the circle, its projection P oscillates back and forth through a middle position C between two extreme positions A and B. The sort of motion described by P along the line AB will now be investigated.

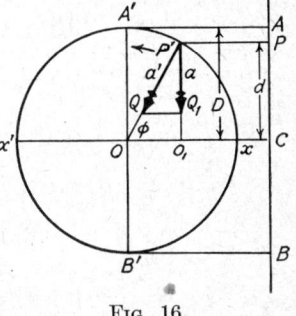

Fig. 16

As the particle P' moves with uniform speed in the circumference of a circle there is a constant acceleration directed toward the center of the circle. In Fig. 16, this radial acceleration a' is represented by the line $P'Q$. Let the component, in the direction of the line AB, of the radial acceleration a' be called a. This a is the acceleration of the point P. Let d represent the displacement of P from the point C. We shall count d positive when above C and negative when below. We shall measure r outward from O.

From the similar triangles $P'QQ_1$ and $P'OO_1$ we have

$$\frac{a}{a'} = \frac{d}{r}$$

Representing the constant linear speed and angular speed of P' by v and w, respectively, we have

$$a' = -\frac{v^2}{r}; \quad \text{and} \quad v = wr$$

Eliminating a' and v from these three equations, we get

$$a = -w^2 d \tag{31}$$

Since w is constant, it follows from this equation that the acceleration of P is proportional to its distance from the center of its path and is opposite in sign. That is, if a point moves with uniform speed in the circumference of a circle, the projection of the point on any straight line in the plane of the circle moves with simple harmonic motion of translation.

19. The Period of a Simple Harmonic Motion of Translation. — The theorem we have just proved will now be used for the determination of the value of the constant c in the defining equation of simple harmonic motion (30).

A comparison of (30) and (31) shows that $c = w^2$. If the time of one revolution of P', that is, the time of one complete vibration of P, be denoted by T, we have

$$w = \frac{2\pi}{T} \text{ radians per unit of time.}$$

$$c\ [= w^2] = \left(\frac{2\pi}{T}\right)^2$$

Hence, if a body of mass m is moving with simple harmonic motion of period T, then when the body is at a distance d from the middle of its path, the acceleration is directed toward the middle of the path and has the value

$$a[= -w^2 d] = -\left(\frac{2\pi}{T}\right)^2 d \tag{32}$$

This body is urged toward the middle of the path by a force

$$F[= ma] = - m\left(\frac{2\pi}{T}\right)^2 d \qquad (33)$$

It follows that the period of a simple harmonic motion of translation is

$$T = 2\pi\sqrt{-\frac{d}{a}} \qquad (34)$$

the minus sign indicating that the displacement and the acceleration are in opposite directions.

20. Simple Harmonic Motion of Rotation. — When a body rotates back and forth with a motion such that the angular acceleration is always directed toward a position of equilibrium and is always proportional to the angular displacement of the body from that position, the body is said to have a *simple harmonic motion of rotation*. The defining equation of simple harmonic motion of rotation is

$$\mathbf{a} = -b\phi \qquad (35)$$

where b is a positive constant.

When a body is vibrating about any axis with simple harmonic motion of rotation, any point on a line fixed in the body, and not on the axis of vibration, moves back and forth in the arc of a circle with a linear acceleration of which the component tangent to the arc has a magnitude which is proportional to the linear displacement of the point, measured along the arc, from the position of equilibrium. Hence the value of the period of a simple harmonic motion of rotation as well as the value of the angular speed, acceleration, and displacement at any time can be obtained from the values of the period, linear speed, acceleration and displacement of a point on a line fixed in the body.

Angular Acceleration and Period. — Substituting in (32) values of linear acceleration and linear displacement in terms of the corresponding angular quantities, (17) and (7), we find that

$$\mathbf{a} = -\left(\frac{2\pi}{T}\right)^2 \phi \qquad (36)$$

Solving for T and substituting for $\mathbf{a}$ its value from (19)

$$T = 2\pi\sqrt{-\frac{\phi}{\mathbf{a}}} = 2\pi\sqrt{-\frac{K\phi}{L}} = 2\pi\sqrt{-\frac{K}{S}} \qquad (37)$$

where L is the torque acting upon a body of moment of inertia K with respect to the axis of vibration, at the instant when the angular displacement of the body from the equilibrium position is ϕ. The symbol S represents the *torsional stiffness*, $[L/\phi]$, of the body.

Angular Velocity. — Figure 17 represents a body X that is vibrating with simple harmonic motion of rotation through an amplitude Φ about an axis normal to the plane of the diagram through the point O. The point p of a line op, fixed in the body, moves back and forth with simple harmonic motion of translation in the arc AB. The distance AB is shown as a straight line $A'B'$

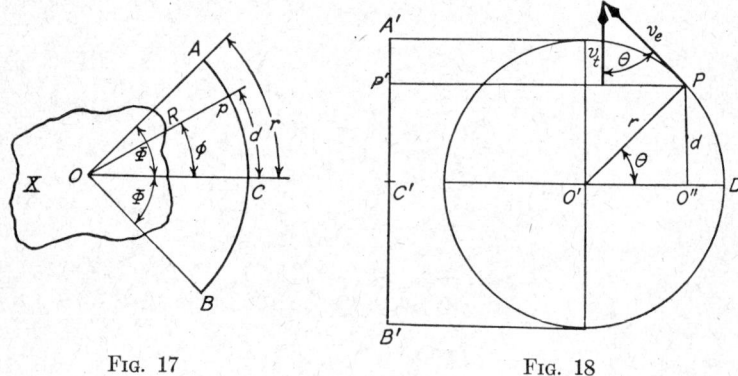

FIG. 17   FIG. 18

in Fig. 18. The simple harmonic motion of translation along $A'B'$ is also the motion of the projection P' of a point P that moves with uniform speed in the circumference of a circle of diameter equal to $A'B'$ and in the same period as the simple harmonic motion of rotation of the body X.

Suppose that the point P moves with uniform speed v_e around the circle and that in time t after passing the position D, the radius $O'P$ has moved through an angle θ. At this instant, the component velocity in the direction $A'B'$, that is, the velocity of the projected point P', has the magnitude

$$v_t = v_e \cos \theta$$

In Fig. 18, the linear velocity of the point P' when at the equilibrium position is v_e, and the value at t seconds later when the body X has rotated through an angle ϕ, is v_t. Dividing each member of this equation by the distance R of the point p from the axis of vibration O,

$$\frac{v_t}{R} = \frac{v_e}{R} \cos \theta$$

or, (16):
$$w_t = w_e \cos \theta \tag{38}$$

SIMPLE HARMONIC MOTION

where w_e and w_t represent, respectively, the angular velocities of the body X when moving through the equilibrium position, and t seconds later.

In Fig. 18 the angle θ equals the product of the angular velocity of the line $O'P$ and the time t occupied in moving through this angle. If time T is occupied in making one entire revolution about O', then the angular velocity of $O'P$ has the magnitude $2\pi/T$. Hence, the angle

$$\theta = \frac{2\pi t}{T} \text{ radians} \tag{39}$$

and (38) becomes

$$w_t = w_e \cos \frac{2\pi t}{T} \tag{40}$$

In this equation time is reckoned from the equilibrium position. If it be reckoned from one end of the oscillation, that is $\pi/2$ radians from the equilibrium position, we would have

$$w_t' = w_e \cos\left(\frac{2\pi t}{T} - \frac{\pi}{2}\right) = w_e \sin \frac{2\pi t}{T} \tag{41}$$

The maximum angular displacement from the equilibrium position is called the *amplitude* of the vibration. Another useful formula for the angular velocity is one involving the period T, angular amplitude Φ and angular displacement ϕ. From Fig. 18, the linear velocity of a body moving with simple harmonic motion of translation of period T and amplitude r at the instant when the displacement from the equilibrium position is d has the value

$$v_t[= v_e \cos \theta] = \frac{2\pi r}{T}\left(\frac{O'O''}{r}\right) = \frac{2\pi}{T}\sqrt{r^2 - d^2} \tag{42}$$

From Fig. 17, (7), and (16)

$$r = \Phi R, \, d = \phi R \quad \text{and} \quad v_t = w_t R \tag{43}$$

Substituting these values in (42) we find

$$w_t = \frac{2\pi}{T}\sqrt{\Phi^2 - \phi^2} \tag{44}$$

At the equilibrium position, $\phi = 0$. When at this angular displacement, the velocity w_e has the magnitude, (44):

$$w_e = \frac{2\pi}{T}\Phi \text{ radians per sec.} \tag{45}$$

If the angular amplitude, expressed in degrees, be represented by $\Phi°$, (45) gives

$$w_e = \frac{2\pi}{T}\frac{\Phi°}{57.3} \text{ rad. per sec.} = \frac{\Phi°}{9.12\ T} \text{ rad. per sec.} \qquad (46)$$

Angular Displacement. If a body is vibrating with simple harmonic motion of translation of period T and amplitude r, the linear displacement from the equilibrium position t seconds after traversing the equilibrium position has the value, Fig. 18:

$$d = r \sin \theta = r \sin \frac{2\pi t}{T} = r \cos\left(\frac{\pi}{2} - \frac{2\pi t}{T}\right) \qquad (47)$$

From (43), we find that when a body has a simple harmonic motion of rotation with period T and amplitude Φ, the angular displacement from its equilibrium position at time t is given by the expression

$$\phi = \Phi \sin \theta = \Phi \sin \frac{2\pi t}{T} = \Phi \cos\left(\frac{\pi}{2} - \frac{2\pi t}{T}\right) \qquad (48)$$

21. The Physical Pendulum. — A compound or physical pendulum consists of a suspended rigid body free to oscillate about a horizontal axis. Consider the physical pendulum AC (Fig. 19), consisting of a body of mass m supported on an axis normal to the plane of the diagram and passing through the point A. Let the center of gravity of the pendulum be at C. Denote the distance AC by l. If the pendulum be deflected from its equilibrium position through an angle ϕ, it will be acted upon by a torque which tends to restore it to the equilibrium position and which has a value

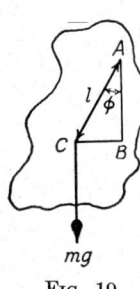

Fig. 19

$$L = -mg(BC) = -mgl \sin \phi$$

the negative sign indicating that the direction of the torque is opposite that of the displacement.

If the displacement from the equilibrium position is small, $\sin \phi$ is approximately equal to ϕ radians. In this case the above equation becomes

$$L \doteq -mgl\ \phi \qquad (49)$$

Whence, at any instant, a pendulum displaced but a small distance from its equilibrium position is urged toward its equilibrium position by a torque nearly proportional to its angular displacement from that position. Consequently, the angular motion of such a pendulum is approximately simple harmonic motion of rotation.

The period of vibration of a physical pendulum oscillating through a small amplitude now will be determined. Substituting in the equation for the period of vibration of a simple harmonic motion of rotation, (37), the value of the torque acting on a compound pendulum displaced through a small angle ϕ from its position of equilibrium, (49), we obtain for the value of the time occupied by one complete vibration of a compound pendulum:

$$T \doteqdot 2\pi\sqrt{\frac{K}{mgl}} \qquad (50)$$

From this equation it is seen that when the amplitude of vibration of a compound pendulum is so small that $\sin \phi$ may be replaced by ϕ, the period of vibration of the pendulum is practically independent of the amplitude of swing.

A heavy particle suspended by a string which is both inextensible and massless, and which is capable of swinging in a vertical plane, is called a *simple* or *mathematical pendulum*. Since the moment of inertia with respect to the axis of oscillation of a simple pendulum of length l and mass m is (21):

$$K = ml^2$$

The period of vibration of a simple pendulum is (50):

$$T_s \doteqdot 2\pi\sqrt{\frac{l}{g}} \qquad (51)$$

A pendulum consisting of a small spherical bob supported by a very thin light inextensible string approximates closely to a simple pendulum.

A rigid body free to oscillate about a horizontal axis that is below the center of mass of the body is called an *inverted pendulum*. One type of seismograph for recording earth tremors consists of an inverted pendulum having the upper end normally held in position by a set of springs.

22. The Conical Pendulum. — If a body, suspended at a point not coincident with the center of mass of the body, be given an impulse directed to one side of the vertical line through the point of support, each line of the body passing through the point of support will sweep out a conical surface. The body will move around and around with a definite period. A suspended body which, owing to its weight, is capable of rotating in such a manner that the line through the point of suspension and the center of mass has constant angular velocity about the vertical line through the point of suspension is called a *conical pendulum*.

An expression for the period of a conical pendulum now is to be determined. In Fig. 20, a body of mass m with center of mass at B is supported without friction at A. When set into motion, the body rotates so that the center of mass traces a circle of radius r about a center C vertically below the point of suspension A.

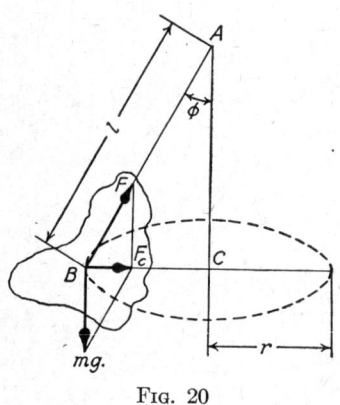

Fig. 20

The resultant force acting on the body is the centripetal force directed toward the center of the circle. This is due to the weight mg of the body, and the reaction F of the support. Representing the distance AB by l, the angle BAC by ϕ, the linear speed of the center of mass in the horizontal circle by v, and the period of a revolution by T, the value of the centripetal force is

$$F_c = [F \sin \phi] = \frac{mv^2}{r} = \frac{m}{r}\left(\frac{2\pi r}{T}\right)^2$$

Also,
$$F \cos \phi = mg$$

Dividing each member of the former equation by the corresponding member of the latter, we have:

$$\frac{\sin \phi}{\cos \phi} = \frac{4\pi^2 r}{T^2 g} = \frac{4\pi^2 l \sin \phi}{T^2 g}$$

Whence, the period
$$T = 2\pi \sqrt{\frac{l \cos \phi}{g}} \qquad (52)$$

23. Wave Motions and Wave Forms. — A motion that goes through the same series of changes at regularly recurring intervals is called *periodic*. A periodic disturbance which is handed on successively from one portion of a medium to another is called a *wave motion*.

It is convenient to represent graphically a periodic function by a curve coördinating the function and time. Thus, if we plot instantaneous values of the electromotive force in an alternating current circuit against time, we obtain what is called the wave form of the periodic alternating current. If we plot a ship's instantaneous angle of roll against time we obtain the wave form of the roll. If the quantity varies as the sine or the cosine of an angle, then the wave form is called a sine curve, or a cosine curve, respectively. Either a sine curve or a cosine curve is a harmonic curve or harmonic wave form.

The value of the angular displacement from the equilibrium position of a body vibrating with simple harmonic motion of rotation is given by (48), where T is the period of the vibration, Φ

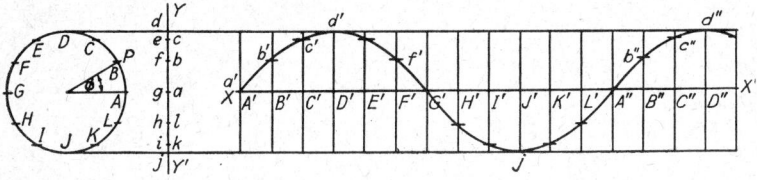

Fig. 21

is the amplitude, and ϕ is the angular displacement from the equilibrium position at time t. The angular displacements made in succeeding equal time intervals are proportional to the distances ab, bc, cd, de, ef, fg, etc., Fig. 21. To obtain these unequal distances, project onto the line YY' the positions at the end of equal time intervals of a point P moving with uniform speed in the circumference of a circle (Art. 18). In the case here represented, the circle, Fig. 21, is divided into twelve equal arcs, AB, BC, CD, etc. On the line XX' mark thirteen points A', B', C', etc., at equal distances apart. Through each of these points draw a line perpendicular to XX'. Draw a line parallel to XX' from each of the points marked on the line YY'. Place a dot a' at the intersection of the horizontal line through a and the vertical line through A'. Similarly, place a dot b' at the intersection of the horizontal line through b and the vertical line through B', and so on for the remainder of the intersections c', d', etc. The smooth curve drawn through these dots is the wave form of the simple harmonic motion represented by (48). The period T is the time required to make one complete vibration. At the time $t = \dfrac{T}{12}$, the angular displacement ϕ is proportional to the distance $B'b'$. The angular amplitude Φ is proportional to the distance $D'd'$.

The wave form due to a simple harmonic vibration is a sine curve (or cosine curve).

24. Phase and Phase Angle. — In the case of vibrations along a horizontal line, forces, velocities, currents, etc., directed to the right are usually termed positive; and displacements to the right of the equilibrium point are termed positive. In the case of vibrations along a vertical line, forces, currents, etc., directed upward are usually termed positive; and displacements above the equilibrium point are called positive. The *phase* of any periodically changing function is the fraction of a whole period of vibration which has elapsed since the particular function last passed the equilibrium position in the positive direction. Thus, if, in Fig. 21, a body vibrates along the line YY' with simple harmonic motion, and if, at intervals of one-twelfth of the period of vibration, the body is successively in the positions $a, b, c, d, e, f, g, h, i, j, k, l, a$, etc., the phase of the displacement of the body when at a is zero; when at c and moving in the positive direction, the phase is $\frac{1}{6}$; when at d, the phase is $\frac{1}{4}$; when at c and moving in the negative direction the phase is $\frac{1}{3}$; when at a and moving in the negative direction it is $\frac{1}{2}$, etc.

If a body is rotating in a circle, the angular displacement of the body from some reference position is called the *phase angle* of the motion. Thus, in Fig. 21, when the point P is at B, the phase of the displacement is $\frac{1}{12}$ and its phase angle is $\dfrac{\pi}{6}$ radians, or $30°$; when at D the phase is $\frac{1}{4}$ and its phase angle is $\dfrac{\pi}{2}$ radians, or $90°$. Since the motion on a straight line of the projection of a point that is moving with uniform speed in the circumference of a circle is a simple harmonic motion of translation, we may express phase in terms of the phase angle of a point that is moving in the circumference of a circle. Thus, in Fig. 21, when the body is at b, and moving in the positive direction, the phase of the motion is $\frac{1}{12}$ and its phase angle is $\dfrac{\pi}{6}$ radians or $30°$. In the equations of Art. 20, the angle $\theta \left[= \dfrac{2\pi t}{T} \right]$ represents the phase angle of the vibration at the time t.

SIMPLE HARMONIC MOTION

Two motions having phase angles that differ by zero or by any integral multiple of 2π radians are said to be in the same phase. Two motions having phase angles that differ by π radians are said to be in opposite phases. In Fig. 21, the properties corresponding to b' and b'' are in the same phase, and those corresponding to d' and d'' are in the same phase. The properties corresponding to d' and j are in opposite phase.

25. The Mean Value of the Product of Two Simple Harmonic Functions of Equal Period. — There are many cases in which it is necessary to know the mean value of the product of two quantities that are varying periodically. If, at any instant, a body rotating with an angular speed w exerts a torque L, the power developed at that instant equals wL. If the torque and the angular speed are varying periodically, the mean power developed during one cycle equals the average of a series of products of w and L during the cycle.

Similarly, the mean power associated with an alternating electric current during one cycle equals the average of a series of products of instantaneous values of electromotive force and current during the cycle.

The most important case is that in which the two periodic quantities vary harmonically and are of the same period. Assume two simple harmonic motions represented, respectively, by the equations

$$d_1 = r_1 \sin \theta \quad \text{and} \quad d_2 = r_2 \sin (\theta + \beta)$$

where r_1 and r_2 are the amplitudes of vibration, d_1 and d_2 are the displacements from the equilibrium position, and θ and $(\theta + \beta)$

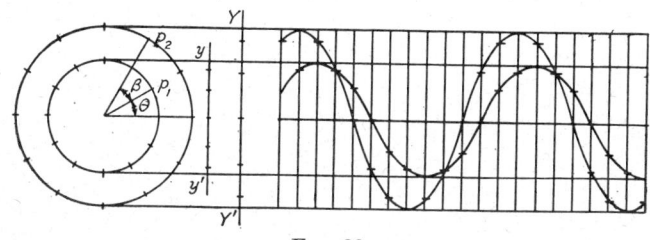

Fig. 22

are the phase angles at the time t. Since the periods are the same, β is independent of time. In Fig. 22, the displacements of the two vibrations at the end of a series of equal time intervals are indicated by dots on the lines yy' and YY', respectively. The wave forms of the two vibrations are sine curves as shown. The product of the two functions at any particular instant is*

$$d_1 d_2 = r_1 r_2 \sin \theta \sin (\theta + \beta) = \frac{r_1 r_2}{2} [\cos \beta - \cos (2\theta + \beta)] \quad (53)$$

Since

$$2\theta + \beta = 2\left(\frac{2\pi t}{T}\right) + \beta$$

it follows that during one-half cycle, that is, while t changes to $t + \frac{1}{2} T$, and θ becomes $(\theta + \pi)$, the angle $(2\theta + \beta)$ becomes $2\left(\frac{2\pi t}{T} + \pi\right) + \beta$. Whence, the change of argument of the last cosine in (53) is 2π radians. The argument of the cosine is a linear function of time. During this change of angle, the cosine goes through all values, positive and negative, and the mean value of $\cos (2\theta + \beta)$ during any half period is zero. Consequently, *the mean value of the product of two simple harmonic functions of equal period and of amplitudes r_1 and r_2 is*

$$\overline{d_1 d_2} = \tfrac{1}{2} r_1 r_2 \cos \beta \quad (54)$$

where β is the difference in phase of the two functions.

For example, if the two harmonically varying quantities d_1 and d_2 represent instantaneous values of electric current and electromotive force, respectively, then the power of an alternating current during any half period is given by (54), where r_1 and r_2 represent the maximum values of the current and electromotive force, respectively, during the half period, and β represents the difference of phase between current and electromotive force. Or, if the two quantities d_1 and d_2 represent instantaneous values of angular speed and torque, respectively, then the power transmitted to the oscillating body during any integral number of half periods is given by (54) where r_1 and r_2 represent the maximum values of the angular speed and torque, respectively, and β represents the difference of phase between angular speed and torque.

Equation (54) shows that:

(a) When two harmonic functions, d_1 and d_2, of equal period are in the same phase, that is, when $\beta = 0$, the mean value of the product $\overline{d_1 d_2}$ is maximum and equals one-half of the product of the amplitudes of the two components.

* From the trigonometric relation:
$$\sin x \sin y = \tfrac{1}{2} [\cos (x - y) - \cos (x + y)]$$

(b) When the two components are in quadrature, that is, when the phase difference β equals 90° or $\frac{\pi}{2}$ radians, the mean value of the product $\overline{d_1 d_2}$ equals zero.

(c) When $\beta = \pi$ radians or 180°, the mean value of the product $\overline{d_1 d_2}$ equals one-half of the product of the amplitudes of the two components but with the negative sign.

These three cases are represented graphically in Figs. 23, 24 and

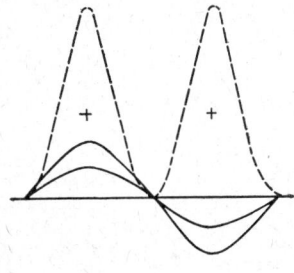

Fig. 23

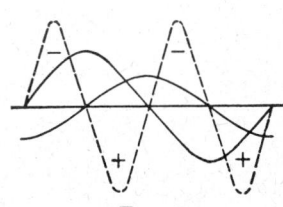

Fig. 24

25, respectively. In each figure, the harmonic curves shown in full lines represent the equations

$$d_1 = r_1 \sin \frac{2\pi t}{T}$$

and

$$d_2 = r_2 \sin \left(\frac{2\pi t}{T} + \beta \right)$$

where β is the difference in phase. In Fig. 23, the difference in phase $\beta = 0$; in Fig. 24, $\beta = \frac{\pi}{2}$ radians or 90°; in Fig. 25, $\beta = \pi$ radians or 180°. In each figure, the ordinate of the dashed curve corresponding to any particular instant equals the product of the ordinates of the two harmonic curves at that instant. Thus, the dashed curves represent the variation of the product, at every instant, of d_1 and d_2. In case d_1 and d_2 represent either electromotive force and current or torque and angular velocity, then the product is power. The dashed curve then shows the variation of power with respect to time. The area between

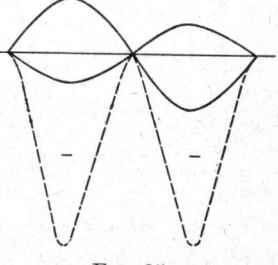

Fig. 25

the power curve and the axis, for any time interval, is proportional to the work done during the given interval. An area above the axis represents positive work, that is work done by the electromotive force or by the torque; an area below the axis represents negative work, that is work done against the electromotive force or against the applied torque. When the two harmonic curves are of equal period and in quadrature, that is, when they differ in phase by $\frac{\pi}{2}$ radians or 90°, Fig. 24, the total work done during one cycle is zero.

The quotient obtained by dividing the area under the power curve for one-half cycle, by the distance between the intersections of this curve with the axis of the curve, equals the mean value of the power during the half cycle. When the two curves are in quadrature, Fig. 24, the mean power during a complete cycle is zero.

The phase difference between the current and the electromotive force in an alternating current circuit can be altered by changing either the capacitance or the inductance of the circuit. If the current and the electromotive force in a circuit containing an alternating current generator and motor are in the same phase, then the generator is putting electric energy into the line and the motor is transforming a maximum fraction of it into mechanical work. If the current and electromotive force differ in phase by one-quarter period, then we have a "wattless current" and the work done during one complete cycle is zero. If the current and the electromotive force are in opposite phase, then the motor is operating as a generator, opposing the electromotive force of the generator.

26. Oscillations of a Coupled System. Resonance. — The vibrations executed by a body which has been displaced from its position of equilibrium and then released are called *free* vibrations. The vibrations executed under the action of an external periodically varying force are called *forced* vibrations.

If the point of support of a vibrating pendulum is in periodic motion, as when a pendulum is attached to another pendulum, or when a pendulum is on a ship which is rolling with a constant period, the pendulum will acquire a compound periodic motion having two components of different periods. These periods will be different from the periods of the freely vibrating pendulum or of the freely vibrating support, except when the inertia of the pendulum is negligible compared with the inertia of the support.

SIMPLE HARMONIC MOTION

The amplitude of the component oscillation of shorter period will be smaller than the amplitude of the component oscillation of longer period.*

Consider a group of pendulums hung from a horizontal rod supported as shown in Fig. 26. Suppose that the pendulum A is set swinging. When it is out on one side of its path it urges the supporting rod out toward that same side. When it is out on the other side of its path it urges the support out toward that side. The force acting on the support varies harmonically. Soon the support and the other pendulums are set into forced vibration of nearly the same frequency as the free vibration frequency of the exciting pendulum A.

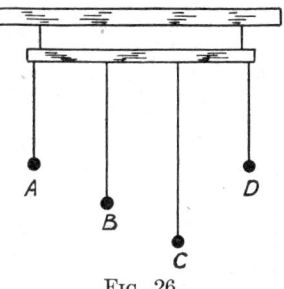

Fig. 26

If the length of the pendulum A is so much greater than the lengths of the other pendulums that the frequency of the forced vibration is much less than the natural vibration frequency of these pendulums when vibrating freely, then they will vibrate with the frequency of the forced vibration. If the frequency of the forced vibration be increased by shortening the length of the exciting pendulum A, for example, the amplitude of the vibration of the other pendulums will increase. If the frequency of the forced vibration equals the natural frequency of some pendulum D, then this pendulum will be acted upon by the maximum force in the direction of its motion when it is moving through its equilibrium position. Consequently, the amplitude of vibration of D will become large. The vibration produced when a periodically varying force acts upon a body of nearly the same frequency of vibration is called *resonant* forced vibration. The phenomenon of the production of resonant forced vibrations is called *resonance*. The body that excites the resonance is called the exciter, and the one that is set into resonant forced vibration is called the resonator. The frequency of the external force that produces maximum resonance is called the critical frequency.

Resonance is a very important phenomenon in mechanics as well as in electricity, sound and light. Some of the more important aspects of the phenomenon that relate to mechanics are broadly indicated in the present considerations.

* Edser, General Physics for Students, pp. 154–167.

The vibrations forced on the resonator lag behind the exciting force. When the frequencies of the exciter and the resonator are equal, the phase of the vibrations of the resonator are one-quarter period behind the vibrations of the exciter, and the amplitudes of the vibrations of the resonator are maximum. That is the resonator is vibrating through the maximum amplitude when the exciting force is zero. When the frequencies of the two vibrations are nearly equal, a small change in the frequency of either produces a considerable variation in the phase difference unless the damping of the resonant vibrations is great.*

Energy is alternately introduced into and withdrawn from the resonator by the exciter but the energy absorbed by the resonator is greater than the energy emitted. The energy absorbed is maximum when the frequency of the resonator is the same as that of the exciter.

Consider the case in which the natural frequency of vibration of D and the frequency of the forced vibration impressed upon it are nearly equal. Suppose that one frequency is n_1 and the other is n_2. Then, during this time interval, the two vibrations are in phase $(n_1 - n_2)$ times. That is, during this time interval, the amplitude of vibration of the resultant motion will rise to a maximum and fall to a minimum $(n_1 - n_2)$ times. This is called the phenomenon of *beats*. If the pendulum A is set into vibration when the periods of A and D are slightly different, then the amplitude of vibration of D will increase to a maximum, fall to a minimum, and repeat. As the energy increasing the amplitude of the vibrations of D comes from the pendulum A, the amplitude of the vibrations of the latter is minimum when the amplitude of D is maximum. Energy of D is now imparted to A thereby increasing the amplitude of its vibrations. Often the frequency of a ship's roll is nearly equal to the frequency of the waves. When this occurs, the amplitudes of successive rolls will become larger and larger to a maximum, thereafter become smaller and smaller to a minimum, and then repeat. As the frequency with which the waves go under the ship depends upon the speed and course of the ship relative to the direction of the waves, the amplitude of roll can be altered by changing either the speed or the direction of the ship.

When the frequency of the forced vibration is much greater than the natural frequency of the body upon which it is impressed,

* Timoshenko, Vibration Problems in Engineering.

the effect upon the latter is very small. A seismograph for recording vibrations of the earth has a pendulum of such small frequency compared to the frequency of the tremors to be measured that it is not set into motion by rapid earthquake vibrations. The instrument records displacements of the earth relative to a pendulum of great moment of inertia and long period.

In this Article the case has been considered in which the periodic force varies harmonically, that is, in which the force is always directed toward the equilibrium position of the body and has a magnitude directly proportional to the displacement of the body from the equilibrium position. It should be noted, however, that if, instead of varying harmonically, the external force starts and stops suddenly but in a periodic manner, it will also set into resonant vibration a body of a free period that is a submultiple of the period of the external force. In this Article it has been assumed that the frequency of the exciter is unaffected by the resonator. The frequency of the exciter will be somewhat modified if either the stiffness or mass of the resonator be not less than that for the exciter.

27. Damping of Vibrations. — When linear vibrations are opposed by a force, or angular vibrations are opposed by a torque, the amplitude of each swing becomes less than the amplitude of the preceding swing. The successive diminution of the amplitudes of swing is called *damping*. If the body vibrating with simple harmonic motion of rotation be acted upon by a damping torque which at each instant is proportional to the angular speed of the body at that instant, then at all ordinary speeds the ratio of the amplitudes of any two successive swings will be constant. In this case the damping is constant.

Figures 27 to 31 represent a pendulous system consisting of a weighted bicycle wheel to which is rigidly attached a pair of connected tanks and partly filled with water or other liquid. The wheel is capable of rotation with negligible friction about a horizontal axle through the center. To produce a pendulum of long period, the moment of inertia with respect to the axis of oscillation is made large, and the torque about the axis of oscillation is made small, (50). In the present piece of apparatus, the moment of inertia is made large by replacing the ordinary tire of the bicycle wheel by a coil of wire of some twenty-five pounds. A small torque is produced by attaching a mass of a few ounces at a short distance below the axis of oscillation.

If the pendulous system be tilted, and held in this tilted position, liquid will flow from the higher tank to the lower, and thereafter will flow back and forth several times with a definite period. If the pendulous system be set into oscillation, and the frequency of this oscillation be nearly the same as that of the liquid, then by the time that the system is in its equilibrium position, there will be an excess of liquid in the reservoir which was the lower when the system was displaced from the equilibrium position. Thus, there will be a difference in phase between the oscillation of the pendulous system and the oscillation of the liquid in the tanks.

Fig. 27

The period of oscillation of the liquid, as well as the phase difference between the oscillations of the liquid and of the pendulous system, can be regulated by adjusting a valve V in either of the tubes that connect the two tanks. If the period of flow of liquid back and forth equals the period of the pendulous system, and if the velocity of the liquid is a quarter period ahead of that of the pendulous system, then the following effects will occur.

Consider the action at an instant when the pendulous system has a large deflection, and there are equal amounts of liquid in the two tanks, Fig. 28. While the pendulous system is moving from this position to the equilibrium position, an excess of liquid rises in the tank that was the lower, the excess being maximum when the pendulous system is in the equilibrium position, Fig. 29. During this quarter period, the distance between the center of mass of the pendulous system and the vertical line through the axis of vibration is being diminished. Hence, the torque urging the pendulous system toward the equilibrium position is being diminished faster than it would be if the liquid were not flowing.

While the pendulous system is moving from the equilibrium position, Fig. 29, to the end of the swing, Fig. 30, the excess of liquid in the tank on the advancing side of the pendulous system produces a torque that opposes the motion of the pendulous system.

By the time that the pendulous system again has attained its equilibrium position, a maximum excess of liquid has flowed into the tank A, Fig. 31, which was the lower in Fig. 30. During motion, the torque urging the pendulous system towards the equilibrium position is being diminished faster than it would be if the

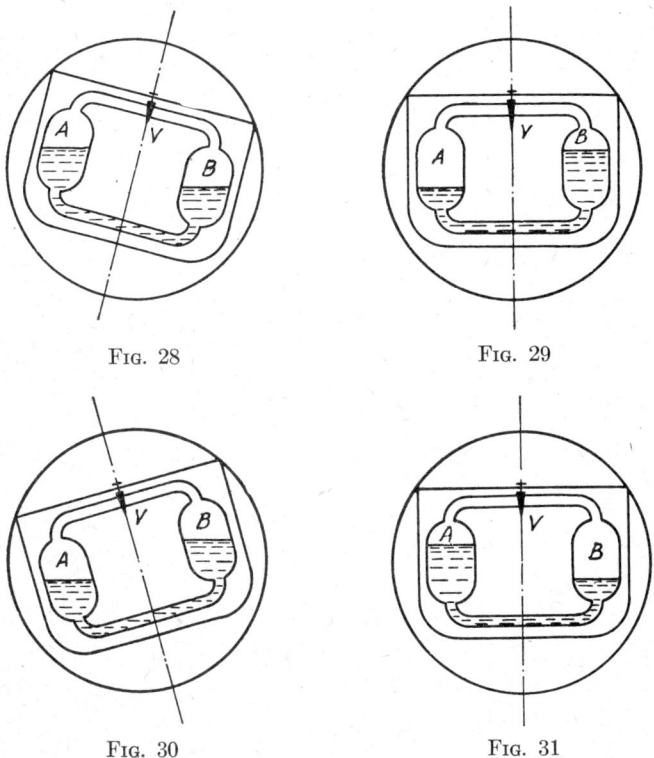

Fig. 28 Fig. 29

Fig. 30 Fig. 31

liquid were not flowing. Consequently, the amplitude of swing is less than it would be if the liquid were not flowing.

Thus, the amplitude of vibration of the pendulous system is being damped throughout the half cycle while the system is moving from its equilibrium position to the end of its path and back. During the next half cycle the amplitude of vibration is damped in the same way. The damping of the amplitude of vibration involves abstraction of energy from the vibrating pendulous system. This energy is transformed into heat as the liquid and air surge back and forth through the constricted passages connecting the two tanks.

For the maximum rate of damping, the periods of the pendulous system and of the oscillating liquid should be equal, and the angular velocity of the pendulum and the torque acting on the pendulum due to the oscillating liquid should differ in phase by one-half period (Art. 25). These factors can be controlled within certain limits by the valve V.

This method has been used for damping the roll of ships (Art. 28), and is used for damping the vibration of certain gyro-compasses (Art. 137).

28. The Frahm Anti-Roll Tanks. — The rolling of a ship by periodic waves can be diminished through resonance by an oscillating mass of water surging back and forth from a tank on one side of the ship to a tank on the other side. The frequencies of sea waves vary from about five to eight per minute in deep water, and from about eight to eleven in shallow water. The period at any particular place does not remain constant for many minutes in succession. The phase of the wave motion changes whenever a wave crest topples over.

The Frahm anti-rolling device* consists of closed tanks placed opposite to one another at the two sides of the ship. The tanks are about half filled with water. In the earlier equipments designed by Frahm, the tanks were placed inside the ship, the lower parts of opposite tanks were joined by a water passage and the upper parts joined by an air passage controlled by a valve. In later equipments, the tanks have been placed on the exterior of the hull like the "blisters" commonly used on warships as a protection against torpedoes. The Hamburg-America line steamers Albert Ballin, Deutschland, Hamburg and New York are provided with such blisters extending above and below the water line along about two-thirds the length of the hull. The blister on each side is divided into three parts by vertical partitions. The upper parts of opposite tanks are joined by air pipes provided with adjustable valves. The lower parts of opposite tanks are not joined by water pipes but open into the sea, Fig. 32. The period of the water surging back and forth from one side to the other can be adjusted within certain limits by adjusting the air valves.

When the period of the waves is about the same as that of the

* U. S. Patent. Frahm, No. 970368, 1910; Frahm, No. 1007348, 1911.

SIMPLE HARMONIC MOTION

ship, the amplitude of roll of the ship will grow larger and larger through resonance. It is the purpose of the tanks to prevent such a building up of the roll amplitude. Suppose that the period of oscillation of the water back and forth from one tank to the other is about the same as the period of roll of the ship and that the velocity of the water surging back and forth from one tank to the other is one-half period behind the torque acting on the ship due

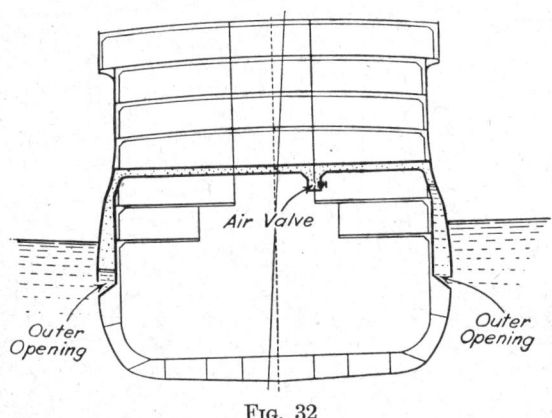

Fig. 32

to the waves. Under these conditions the surging water will absorb energy from the rolling ship at the maximum rate (Art. 25). The required adjustment of period of surge of the water in the tanks and of the phase difference between the angular velocity of the ship and the torque acting on the ship, can be made by a man operating the valve in the pipe connecting the tanks. The maintenance of the proper adjustment is rendered very difficult by the two facts that the phase of the torque acting on the ship is changed when an oncoming wave topples over, and that the period of the waves is constant for but short spaces of time.

It has been proposed to use a gyroscope to operate this valve automatically.[*]

The weight of the Frahm anti-rolling tanks and contents is from about three to five per cent of the weight of the ship, depending upon the constants of the ship.[†]

[*] Hammond, No. 1700406, 1929.
[†] Robb, Studies in Naval Architecture, pp. 289–304.

CHAPTER II

THE MOTION OF A SPINNING BODY UNDER THE ACTION OF A TORQUE

29. Degrees of Freedom. — A locomotive, having a motion limited to motion back and forth along a straight track, is said to have one degree of freedom of translation. A block of ice on the surface of a frozen lake has two degrees of freedom of translation. A bird in the air can move in any direction and is said to have three degrees of freedom of translation or to be unconstrained with respect to linear motion.

A body that is rotating about an axis through the center of mass of the body is said to be *spinning*. The flywheel of a stationary engine is spinning about a fixed horizontal axis and is said to have one degree of freedom of rotation. If the engine were on a turntable capable of rotating about an axis perpendicular to the spin-axis, then the flywheel would be said to have two degrees of freedom of rotation. If the flywheel were mounted so that the spin-axis could rotate about two axes perpendicular to one another and also to the spin-axis, then the wheel would be said to have three degrees of freedom of rotation or to be unconstrained with respect to rotation. A wheel mounted so as to be capable of rotation about five intersecting axes, Fig. 33, has three degrees of freedom of rotation, whatever may be the directions of the axes. A wheel mounted so as to be capable of rotation about three intersecting axes has three degrees of freedom of rotation except when all three axes are in the same plane, Fig. 34. When all three axes of the mounting are in the plane of the spin-axle, the wheel cannot rotate about an axis perpendicular to that plane.

FIG. 33

A ship's compass should stand upright however the ship may either roll or pitch. The same is true of the oil lamps carried by sailing vessels. Each should be free to rotate relative to the ship about an athwartship axis and also about a fore-and-aft axis, that is, should have two degrees of rotational freedom about two perpendicular horizontal axes. This result is attained by supporting the apparatus pendulously by two knife-edges on a ring which in turn is supported on two other knife-edges. The axes of the knife-edges intersect at right angles to one another. Such supporting rings are called *gimbal rings* or *gimbals*. The system of mounting is often called *Cardan's suspension* and is in common use when a device must be supported so as to have two degrees of rotational freedom. The two inner rings of Fig. 34 constitute a Cardan suspension for the gyro-wheel. The gyro has a third degree of rotational freedom about the spin-axis.

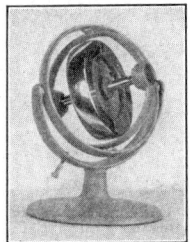

FIG. 34

By means of a rapidly spinning wheel mounted so as to have three degrees of rotational freedom, Foucault in 1852 demonstrated the rotation of the earth. Since this apparatus exhibits the rotation of any body on which it is placed, Foucault called it a "gyroscope." If the direction of the spin-axle is changed, the moment of inertia of the wheel opposes the angular displacement. When it is desired to fix the attention on the property of the spinning wheel to oppose any change in the direction of the spin-axle, the instrument is called a "gyrostat." The spinning body is called the gyro-wheel or *gyro*. That part of physics dealing with the laws of motion of a spinning body under the action of a torque about an axis inclined to the spin-axle is called *gyrodynamics*. Any body or instrument that exhibits the laws of gyrodynamics may be called a *gyroscope*. The instruments represented in Figs. 33 and 34 are called Bohnenberger's gyroscope of five frames, and of three frames, respectively. By means of clamps, rotation of one or more frames can be prevented, thereby reducing the number of degrees of rotational freedom of the gyro to two or one.

30. The Effect on the Motion of a Spinning Body Produced by an External Torque. *Experiment.* — Attach a small mass m to the inner supporting frame of the gyroscope, Fig. 35, at a point near one end of the gyro-axle. Set the gyro-wheel spinning in the direction indicated in the figure by the arrow h_s. At first, the spin-axle will oscillate slightly up and down and also to the right and left. Wait till the motion is steady and then observe that, although the weight of m produces a torque about the axis L, the dip of the gyro-axle about the torque-axis is not conspicuous. Observe that so long as a constant external torque acts about an axis perpendicular to the spin-axle, the spin-axle rotates with constant angular velocity w_p about an axis perpendicular to both the spin-axis and the torque-axis.

An outside torque acting upon an unconstrained spinning symmetrical body about the spin-axis changes the magnitude of the spin-velocity but not the direction of the spin-axis. An outside torque acting about an axis perpendicular to the spin-axis does not change the magnitude of the spin-velocity but does change the direction of the spin-axis in the manner now to be explained. The cause of the oscillations of the spin-axle before a steady state is attained will be considered in Art. 39.

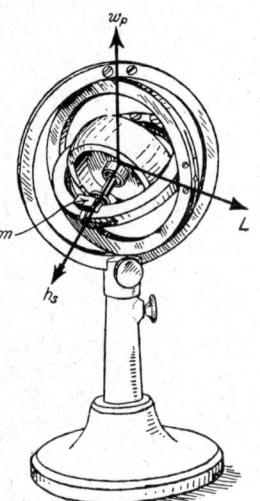

Fig. 35

We shall use two different methods of explanation. In the first method, we shall consider a symmetrical wheel, Fig. 36, spinning with angular velocity w_s about the axle of the wheel, and also rotating with angular velocity w_p about an axis perpendicular to the spin-axis. The resultant of these two angular velocities is an angular velocity represented in magnitude, direction of axis and sense of rotation by the arrow labeled w_r.

Each particle of the wheel is moving in a circular path about some point on the axis of w_r as a center. In order that this motion may continue, each particle must be acted upon by a force directed toward the center of its circular path (Art. 9). The centripetal forces acting on two particles A and A' situated at points equally distant from the axis of the resultant angular velocity w_r, are represented by the arrows f and f'. These two forces constitute a couple acting in the counter-clockwise direction about an axis perpendicular to the axes of w_s and w_p. The sum of the centripetal couples that must act upon all the particles composing the spinning body consti-

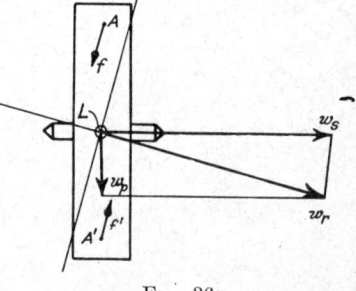

Fig. 36

MOTION OF A SPINNING BODY

tutes a resultant centripetal torque L. A torque of this value must be applied to the spinning body by an outside agent in order that the body may rotate with angular velocity w_p about an axis perpendicular to the spin-axis.

If such a centripetal torque be not applied by some outside agent, the inertia of each particle of the spinning body will cause it to move tangentially to its circular path and the wheel will not turn about any axis except the spin-axis. In case the face of the wheel were not perpendicular to the spin-axis, a torque would act upon the wheel tending to set the face of the wheel perpendicular to the spin-axis. The magnitude of this torque equals the centripetal couple required to hold the wheel inclined to the spin-axis.

Any rotating body tends to set itself so that its axis of maximum moment of inertia is in the direction of the axis of rotation.

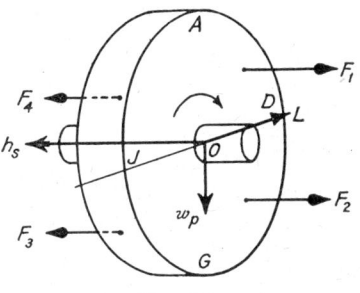

Fig. 37

In the second method of explanation, we shall assume that the wheel in Fig. 37 is spinning at a constant rate in the direction indicated and, at the same time, is turning at a constant rate about the axis JD in the direction of an external torque L of constant magnitude. Owing to the resultant of these two motions, particles in the quadrant OAD and in the quadrant OGJ are approaching the axis JD while particles in the quadrants ODG and OJA are receding from this axis.

When a particle approaches the axis JD, its moment of inertia relative to that axis decreases. Since the rate of spin is constant, the angular speed of each particle about the axis JD increases when the particle is approaching this axis. The angular motion about the axis JD of each particle in the quadrant OAD is accelerating toward the reader about the axis JD, while the motion of each particle in the quadrant OGJ is accelerating away from the reader. Thus, each particle in the quadrant OAD is acted upon by an external force directed toward the reader, while each particle in the quadrant OGJ is acted upon by a force directed away from the reader.

When a particle recedes from the axis JD, its moment of inertia relative to that axis increases. The rate of spin being constant, the angular speed of each particle about the axis JD decreases when the particle is receding from this axis. The angular motion about the axis JD of each particle in the quadrant ODG is decelerating away from the reader (accelerating toward the reader), while the motion of each particle in the quadrant OJA is accelerating away from the reader. Thus, each particle in the quadrant ODG is acted upon by an external force toward the reader, while each particle in the quadrant OJA is acted upon by a force away from the reader.

The resultant of the forces acting on all the particles in the quadrant OAD is represented by the line F_1, Fig. 37. The resultants of the forces acting on the particles in each of the other three quadrants are represented by the lines of equal length, F_2, F_3 and F_4, respectively. These forces tend to turn the spin-axle about an axis AG in the direction indicated by the arrow-head marked w_p.

It has now been shown that *whenever an external torque is applied to a symmetrical unconstrained spinning body, the spin-axis tends to become parallel to the torque-axis, and with the direction of spin in the direction of the torque.*

If the unconstrained spinning body be acted upon simultaneously by more than one torque, then the spin-axis tends to set itself parallel to the resultant torque.

In the case of a wheel mounted so that there is a fixed angle between the spin-axis and the torque-axis, the torque-axis retreats from the advancing spin-axis. The spin-axis continues to move in the direction to make it become parallel to the torque-axis, but it cannot become parallel on account of the rigidity of the connection between the wheel and its supporting frame.

31. Precession. — In the preceding Article it has been shown that when a symmetrical unconstrained spinning body is acted upon by an external torque about an axis perpendicular to the spin-axle, the spin-axle will rotate about an axis perpendicular to the axes of spin and of torque. For example, with the directions of spin and of external torque as indicated by the arrows labeled h_s and L in Fig. 38, there would be a rotation of the spin-axle about the axis AG in the direction indicated by the arrow labeled w_p. It will now be shown that this rotation of the spin-axle develops an internal torque that is in opposition to the applied external torque.

Because of the angular velocity w_p, a particle in the semicircle $ADGO$ is approaching the reader, and a particle in the semicircle $GJAO$ is receding from the reader. A particle at either A or G has zero linear velocity and maximum linear acceleration perpendicular to the face of the wheel. A particle at either D or J has maximum linear velocity and zero linear acceleration perpendicular to the face of the wheel.

While a particle of the wheel is moving from A to D its inertia opposes the increase of its linear velocity toward the reader by an internal force directed away from the reader. This force is maximum when the particle is at A and zero when it is at D. Thus, all particles from A to D are urged away from the reader. As a particle of the wheel goes from D to G, its inertia opposes the diminution of the linear velocity toward the observer. Thus, all particles from D to G are urged toward the reader. As each particle goes from G to J its inertia opposes the increase of linear velocity away from the observer, and hence each particle from G to J is urged toward the observer. As each particle goes from J to A, its inertia opposes the diminution

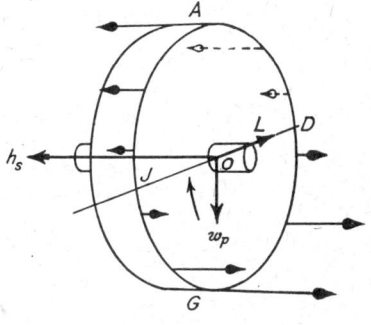

Fig. 38

of linear velocity away from the observer by a force away from the observer. Consequently, the angular velocity about the axis AG develops an internal torque about the L-axis that is in opposition to the external torque about the same axis.

The resisting torque developed when a spinning symmetrical unconstrained body is acted upon by an external torque about an axis perpendicular to the spin-axis, is called *gyroscopic resistance* or *gyroscopic torque*. Gyroscopic torques are internal couples which change the direction of the spin-axle without performing work on the body. The angular motion of the spin-axle accompanied by a resisting torque is called *precession*. The axis about which the spin-axle rotates is called the axis of precession. The angular velocity of the spin-axle about the axis of precession is called the precessional velocity. The precessional velocity is constant when the external torque is of the same magnitude as the internal gyroscopic torque. The external torque that produces or maintains precession is called the *precessional couple* or *precessional torque*.

Questions. 1. A contest sometimes seen in military carnivals is a race in which each contestant rolls a heavy gun wheel around a course. Show that, if the man is running alongside the wheel with the wheel on his right side,

(a) a downward force on the hub will steer the wheel to the left; (b) if the wheel is making a turn to the right and tilting over toward the right, a push forward on the hub will bring the wheel more nearly upright.

2. The face of a certain wheel is not perpendicular to the shaft to which it is keyed. Show that when the shaft rotates it tends to precess and the bearings are acted upon by a torque about an axis which is not constant in direction.

32. Change in the Motion of a Ship Produced by Precession of the Shaft. — Suppose that a wave exerts a torque on a side-wheel steamer causing the steamer to roll to the starboard.* In Fig. 39, the direction of the angular momentum of the paddle-wheels and connecting shaft is indicated by h_s, and the direction of the applied torque by L. The direction of the precessional velocity thereby produced is indicated by w_p. The precession causes the bow to turn to the starboard. The development of the precession w_p causes a torque about the same axis as the precession and in the same sense. Since the spin-axis tends to set itself parallel to this torque-axis, and with the direction of spin in the direction of the torque, it follows that the roll to starboard is opposed. A starboard roll causes the starboard paddle-wheel to dip deeper into the water than the port paddle-wheel. This produces a turning of the bow toward the port, thereby opposing the starboard turning due to precession.

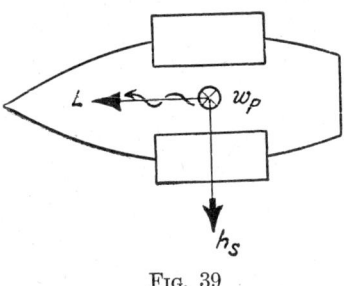

Fig. 39

A side-wheel steamer turning to the starboard under the action of the rudder will heel to the port on account of precession.

In the case of a ship with screw propeller, the effects of rolling and pitching are not the same as in the case of a side-wheel steamer. So long as the shaft is horizontal, a side-wheel steamer is not precessed by pitching and a screw propeller steamer is not precessed by rolling. Suppose that the angular momentum h_s of the shaft and propeller of a ship is in the direction indicated in Fig. 40. If a wave exerts a torque in the direction represented by L tending to raise the bow or lower the stern, the bow of the ship will turn to starboard about a vertical axis with angular velocity

---

* The starboard side of a ship is the right-hand side and the port side is the left-hand side when the observer has his back to the stern and his face toward the bow.

MOTION OF A SPINNING BODY 51

w_p. This precessional velocity implies a torque about the same axis and in the same sense. Since the spin-axis tends to set itself parallel to this torque-axis, and with the direction of spin in the direction of the torque, it follows that the precession decreases the angle of the pitch that produced it.

A side-wheel steamer will roll less and pitch more than a screw propeller steamer if corresponding dimensions are equal.

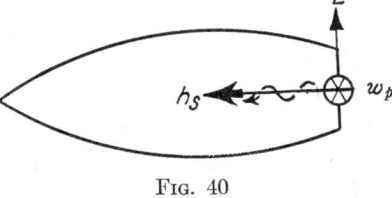

FIG. 40

Show that when a screw propeller steamer is at the same time rolling and pitching, the bow is deflected back and forth out of the proper course of the ship. This effect is called "yawing" or "nosing."*

33. Deviations of the Course of an Airplane Produced by Precession of the Propeller Shaft. — Consider an airplane of which the single propeller, shaft, and rotating parts of the engine have considerable angular momentum h_s with respect to the spin-axis, Fig. 41. If no torque be applied tending to change the direction of the spin-axis, the axle will maintain its direction in space, Art. 16. If a torque be applied in the direction represented by L, by rudders or any other means, tending to tilt the head upward, then the spin-axis with the attached airplane will precess with an angular velocity represented by w_p. With the directions of spin and of torque as indicated in the figure, this precession will turn the head to the right. If the precession be retarded by steering to the left, that is by applying a torque opposite the direction of w_p, then the spin-axis will tend to set itself parallel to the new torque-axis and the head will rise still more. If, however, the precession be accelerated by steering to the right, that is by applying a torque in the direction of w_p, then the head will be tilted downward. Thus it is seen that in order to turn an

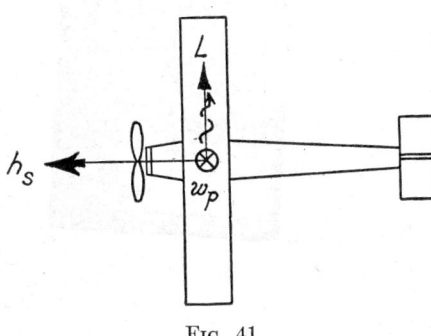

FIG. 41

* Suyehiro, "Yawing of Ships," Trans. Int. Nav. Art., 1920, pp. 93–101.

airplane without causing the machine to tilt, it is necessary to use not only the vertical rudder to produce the turn but also the horizontal rudder to neutralize the tilt that otherwise would be produced by precession. If the airplane has two similar propellers rotating in opposite directions with equal speed, then the precession will be zero.

Questions. 1. Show that in the case of an airplane with a propeller rotating counter-clockwise as viewed by a person looking from aft forward, a turn to the right tends to make the plane soar.
2. Show that an airplane, with a propeller rotating in the clockwise direction as viewed from aft forward, can be turned to the right by elevating a horizontal rudder at the stern.
3. Show that if an airplane makes a sharp turn about a vertical axis and banks properly to furnish the required centripetal force, the airplane will precess about a horizontal athwartship axis causing the nose to dip, even though the engine is not running. Aviators sometimes ascribe this phenomenon to a " hole in the air."
4. In what direction would a horizontal rudder at the stern of an airplane need to be tilted so as to prevent the airplane turning about an athwartship axis while making a turn to the right, the rotation of the propeller being clockwise?
5. In what direction would a horizontal rudder at the stern need to be tilted in order to make a turn to the right without using the vertical rudder, the rotation of the propeller being clockwise?
6. Would the trailing edge of a vertical rudder need to be moved to port or to starboard in order to make the bow of the airplane rise, the rotation of the propeller being clockwise?
7. The Pitcairn-Cierva Autogiro is an airplane to which is added a set of long vanes carried by a hub fastened rigidly to the airplane so that the vanes can revolve above the cockpit about a vertical axis. Each vane is pivoted to the hub in such a way that it can set itself at an angle of 90° or less to the axis of rotation. In starting the airplane, both the propeller and the hub are rotated by the engine in the clockwise direction as viewed by the pilot who is behind the propeller and below the hub. As soon as the airplane is free of the ground, the hub is disconnected from the engine. The motion of the airplane through the air thereafter causes the vanes to rotate about the vertical axis of the hub and also to rise like the ribs of an umbrella turned inside out.

Deduce the direction of any precession of the airplane while (*a*) the propeller is being accelerated; (*b*) the right wing is being lifted by a gust of wind.

34. Magnitude of the Torque Required to Maintain a Given Precessional Velocity when there is Zero Motion about the Torque-Axis, and when the Axes of Spin, of Torque and of Precession are Perpendicular to One Another. *Experiment.* — Stand on a rotatable stool while holding in the hands a bicycle wheel provided with

a massive tire of lead or iron wire. While the wheel is spinning with the axle horizontal, incline the right-hand end of the axle downward, then upward. Observe the considerable torque required to tilt the axle. Also observe that the rotatable stool turns about a vertical axis in one direction when the right-hand end of the axle is being tilted downward and turns in the opposite direction when the same end of the axle is being tilted upward.

Experiment. — When a gyro-wheel of the gyroscope in Fig. 42 is not spinning, push downward against the point x with the rubber-tipped end of a lead pencil. Observe the effect and the amount of push exerted.

Fig. 42

Fasten together the two outer rings of the gyroscope by means of a clamp C, Fig. 43. Set the gyro-wheel spinning. Push downward against the point x as before and observe the effect and the amount of push exerted.

Detach the clamp so that the gyro-axle can precess. Set the gyro-wheel spinning. Push downward against the point x as before and again observe the effect and the amount of push exerted.

Fig. 43

Attach the small mass furnished to the frame at x, Fig. 43. With the clamp C removed, set the wheel spinning in the direction indicated. Observe now that the torque due to the weight of the added mass produces a precessional motion about a vertical axis but that it produces inappreciable motion of the spin-axle about the axis of the torque.

Observe that after the precession has become uniform, the angular momentum produced by a constant torque about an axis perpendicular to the spin-axle does not increase the total angular momentum of the wheel but it changes the direction of the axis of angular momentum. This is analogous to the fact that centripetal force does not increase the total linear momentum of a body traveling in a circular arc but it changes the direction of the linear momentum.

We shall now deduce the relation between the precessional velocity, the angular momentum about the spin-axis, and the torque acting on a body that is symmetrical with respect to both the spin-axis and the precession-axis. Represent the magnitude of the spin-velocity by w_s, the moment of inertia of the wheel with respect to the spin-axis by K_s, and the torque due to the added weight by L.

In Fig. 44, the angular momentum of the gyro-wheel about the spin-axis at one instant is represented by the line S_1O, and at a short interval of time Δt afterward by the line S_2O. During this time interval there has been a small change of angular momentum represented by the line S_2S_1 perpendicular to the spin-axis.

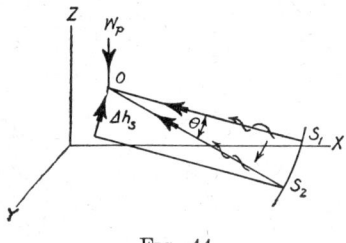

Fig. 44

The time rate of change of angular momentum, that is the torque required to maintain the constant precessional velocity w_p of a body spinning with angular velocity w_s about an axis perpendicular to the spin-axis and to the axis of precession, has the value (28):

$$L = \frac{S_2S_1}{\Delta t}$$

Now during the short time interval Δt, the vertical plane through the spin-axis has rotated about a vertical axis through an angle θ with a constant precessional velocity w_p. Consequently, $\theta = w_p \, \Delta t$. For a sufficiently short time interval, the chord S_2S_1 is approximately equal to the arc S_2S_1. Consequently,

$$\theta \left[= \frac{S_2S_1}{S_2O} \right] = w_p \, \Delta t \quad \text{or} \quad S_2S_1 = (S_2O)w_p \, \Delta t$$

Substituting this value of S_2S_1 in the equation of torque,

$$L \left[= \frac{S_2S_1}{\Delta t} \right] = (S_2O)w_p$$

Remembering that the angular momentum represented by S_2O has the magnitude $K_sw_s[= h_s]$, we find that the torque required to maintain a constant velocity of precession about an axis perpendicular to the spin-axis has the value

$$L[= (S_2O)w_p] = K_sw_sw_p = h_sw_p \qquad (55)$$

where $h_s[= K_sw_s]$ is the angular momentum of the body relative to the spin-axis.

MOTION OF A SPINNING BODY 55

The value of the torque required to maintain constant precession about an axis that is not perpendicular to the spin-axis is deduced in Art. 53.

Since for every torque there is a reacting torque of the same magnitude about the same axis acting upon a different body, it follows that if a body spinning with an angular speed w_s rotates about another axis with angular speed w_p there is developed a reacting torque acting on the restraints having a value L given by (55). The opposition which the axle of a rotating wheel offers to being turned about the axis of an applied torque is called the *gyroscopic couple, gyroscopic resistance,* or the *stabilizing property of the gyroscope.*

From (55) it is seen that

(*a*) If the torque L be zero, then the precession will be zero. For example, if a wheel is mounted so as to be free to move about any axis passing through its center of mass, then the spin-axle remains fixed in space however the frame may be displaced. This is the *First Law of Gyrodynamics.*

(*b*) When a constant torque L is acting on an unconstrained symmetrical spinning body about an axis perpendicular to the spin-axle, the spin-axis will precess steadily about an axis perpendicular to both the spin-axle and the torque-axis with an angular velocity

$$w_p \left[= \frac{L}{h_s} \right] = \frac{L}{K_s w_s} \tag{57}$$

This is the *Second Law of Gyrodynamics.*

(*c*) When precession is prevented, that is, when $w_p = 0$, then the torque $L = 0$. Hence, a rotating body offers zero opposition to a rotation of the axis of spin when precession is prevented.

35. The torque L produces an angular displacement about an axis perpendicular to the torque-axis, but may produce zero displacement about its own axis. Under these conditions, the torque does zero work. This is similar to the fact that a centripetal force acting perpendicular to the direction of the motion of a body moving in a circular path does no work on the body.

Rotating dynamo or motor armatures, and the rotors of steam turbines, mounted lengthwise of a ship, develop on the ship considerable torques when the ship pitches, but not any when the ship rolls. If the axes of rotation be athwartship, there will be torques when the ship rolls, but not any when it pitches. With the mounting in either direction, there will be a torque when the ship moves in a curved course.

In the case of a locomotive rounding a curve, the change in direction of the axles of the drive wheels develops a torque tending to lift the drivers from the inside rail. When a side-wheel steamer turns there is a similar torque. When a rapidly moving motorcycle makes a sharp turn, the axle of the flywheel develops such a large gyroscopic couple that the rider is compelled to lean much more toward the inside of the curve than he would if the machine had no flywheel.

It should be kept in mind that (55) and all equations derived from it apply only to bodies that are symmetrical with respect to the spin-axis, that is, to bodies for which the moments of inertia about all axes through, and perpendicular to, the spin-axis are equal. It applies to rifled projectiles, electric generator and motor armatures, pulleys, flywheels and turbine runners.

This equation requires modification to make it applicable to an unsymmetrical body, that is, one for which the moments of inertia about all axes through, and perpendicular to, the spin-axis are not equal. For such a body another term must be added to (55) of the form $(K_1 - K_2)w_s w_p$ where K_1 and K_2 represent the moments of inertia with respect to two perpendicular axes that intersect the spin-axis at right angles. In the case of an airplane propeller of two blades, for example, the torque L is not uniform throughout a revolution. A scheme to prevent vibration of the machine being developed by this variable torque consists in connecting the hub of the propeller to a fork on the end of the engine shaft in such a manner that the hub is free to rock about a pin perpendicular to the shaft and inclined to the long axis of the propeller at an angle of about 45 degrees.*

The unique precessional velocity given by (57) is for the case in which the precessional axis is perpendicular to the spin-axis. It can be shown that in the case of a top spinning about an axis inclined to the vertical, there are two possible values of the precessional velocity, depending upon the magnitude of the spin-velocity. When the spin-velocity exceeds a certain value, the rapid precession is produced. Under this condition an increase of the external torque applied to the spinning body results in a diminution of the precessional velocity.†

* U. S. Patent. Messick, No. 1491997, 1924.
† Deimel, Mechanics of the Gyroscope, p. 89.

MOTION OF A SPINNING BODY

Problem. A certain motorcycle weighing 300 lb. has two wheels each of 15 lb., radius of gyration 12 in., and outside diameter 28 in. It has two disk-shaped flywheels each of 20 lb. and diameter 10 in. rotating with the same angular velocity as the wheels. The engine makes 3000 r.p.m. when the machine is traveling 60 mi. per hr. The rider weighs 170 lb. The center of mass of the machine is 16 in. above the ground and that of the rider is 36 in. above the ground.

Find the angle through which the machine is tilted from the vertical when the machine has a speed of 60 mi. per hr. around a curve of 300-ft. radius due (*a*) to centripetal effect alone; (*b*) to gyroscopic effect alone.

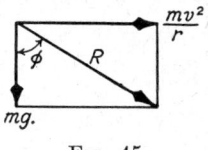

FIG. 45

Solution. 60 mi. per hr. = 88 ft. per sec.

From Fig. 45,

(a) $$\tan \phi = \frac{\frac{mv^2}{r}}{mg} = \frac{v^2}{rg} = \frac{(88)^2}{300 \times 32.1} = 0.8$$

Therefore $\phi = 38°\ 40'$ inclination due to centripetal effect.

(b) The inclination θ due to gyroscopic effect is given by the equation

$$\tan \theta = \frac{\text{gyroscopic force acting horizontally at the center of mass of the system}}{\text{weight of the system}}$$

$$= \frac{\frac{\text{gyroscopic torque}}{\text{height of center of mass above ground}}}{\text{weight of the system}}$$

The gyroscopic torque is given by the equation

$$L = K_s w_s w_p + K_s' w_s' w_p'$$

where the unprimed quantities refer to the motorcycle wheels and the primed quantities refer to the flywheels. From (21)

$$K_s\,[=2\,mk^2] = \frac{2(15)1^2}{32.1} = 0.94 \text{ slug-ft.}^2$$

From (22)

$$K'\left[=\frac{2\,mr^2}{2}\right] = \frac{2(20)5^2}{2 \times 32.1(12)^2} = 0.11 \text{ slug-ft.}^2$$

Again,

$$w_s = \frac{\text{speed of machine}}{\text{radius of wheel}} = \frac{88}{\frac{14}{12}} = 75.6 \text{ rad. per sec.}$$

$$w_s' = \frac{2\pi\,3000}{60} = 314 \text{ rad. per sec.}$$

$$w_p = \frac{\text{speed of machine}}{\text{radius of curve}} = \frac{88}{300} = 0.29 \text{ rad. per sec.}$$

$$w_p' = 0.29 \text{ rad. per sec.}$$

MOTION OF A SPINNING BODY

Consequently, the total gyroscopic torque

$$L[= K_s w_s w_p + K_s' w_s' w_p'] = 0.94(75.6)0.29 + 0.11(314)0.29 = 30.6 \text{ slug-ft.}^2$$

The height X of the center of mass of the man and machine above the ground is, (29),

$$X = \frac{[170(36) + 300(16)]\cos\theta}{170 + 300} = 23.2 \cos\theta \text{ in.} = 1.93 \cos\theta \text{ ft.}$$

Substituting these values in the equation for $\tan\theta$,

$$\tan\theta = \frac{\dfrac{30.6 \text{ slug-ft.}^2}{1.93 \cos\theta \text{ ft.}}}{470 \text{ lb. wt.}} = \frac{0.03}{\cos\theta}$$

Whence the inclination to the vertical due to gyroscopic effect is

$$\theta = \sin^{-1} 0.3 = 1° 40'$$

Problem. Attached to the motorcycle of the preceding problem is a side car weighing 200 lb. and having the center of mass 30 in. from the cycle and 16 in. above the ground. Considering both centripetal and gyroscopic effects, find with what speed the machine must be moving around a curve of 100-ft. radius in order that the sidecar may be on the verge of rising off the ground.

Solution. When the motor and sidecar go around a curve, the system is acted upon by a torque tending to raise the sidecar. This torque is due to the resultant effect of the centripetal force pulling the car out of a straight path and the gyroscopic forces due to the change in direction of the spin-axes of the two flywheels and the three wheels that roll along the ground. The resultant torque tending to raise the sidecar is opposed by a torque due to the weight of the car. Each of these torques is equivalent to the product of a horizontal force along the radius of the circular path and the height of the center of mass of the system above the ground. The horizontal force applied at the center of mass of the entire moving system that is required to raise the wheel of the sidecar off the ground equals the sum of the centripetal force and the gyroscopic force due to the three wheels on the ground and the two flywheels. We shall now find the value of these four quantities and substitute them in the numerator of the second member of the equation,

$$\text{The horizontal counteracting force} = \frac{\text{the force required to raise the sidecar}}{\text{distance of the centroid above the ground}}$$

The torque, $L\left[= \dfrac{30 \times 200}{12} \right] = 500$ lb.-ft.

The distance X above the ground of the center of mass of the system consisting of man, motorcycle and sidecar is, (29),

$$X\left[= \frac{170 \times 36 + 300 \times 16 + 200 \times 16}{670} \right] = 21.1 \text{ in.} = 1.76 \text{ ft.}$$

Whence, the horizontal counteracting force

$$F_h\left[= \frac{L}{X} \right] = \frac{500 \text{ lb.-ft.}}{1.76 \text{ ft.}} = 284 \text{ lb. wt.}$$

MOTION OF A SPINNING BODY

The centripetal force

$$F_c \left[= \frac{mv^2}{r} \right] = \frac{670 \, v^2}{32.1(100)} = 0.208 \, v^2 \text{ lb. wt.}$$

The gyroscopic force due to the three wheels rolling along the ground,

$$F_g \left[= \frac{K_s w_s w_p}{X} \right] = \frac{\frac{3(15)1^2}{32.1} \frac{v}{1.165} \frac{v}{100}}{1.76} = 0.0068 \, v^2 \text{ lb. wt.}$$

The gyroscopic force due to the two flywheels

$$F_g' \left[= \frac{K_s' w_s' w_p'}{X} \right] = \frac{0.11 \left(\frac{2 \pi \, 3000 \, v}{60 \times 88} \right) \frac{v}{100}}{1.76} = 0.0022 \, v^2 \text{ lb. wt.}$$

Substituting these values of F_h, F_e, F_g and F_g' in the equation stated in words at the beginning of this solution,

$$284 = (0.208 + 0.0068 + 0.0022)v^2$$

Whence, the speed with which the machine must move around the given curve in order that the sidecar may be on the verge of rising off the ground is

$$v = 36.15 \text{ ft. per sec.} = 24.6 \text{ mi. per hr.}$$

Problem. A rotary airplane engine of 325 lb., and radius of gyration 14 in., makes 1300 r.p.m. when the plane has a speed of 150 mi. per hr. Find: (a) the gyroscopic torque exerted on the plane when making a turn of 200-ft. radius; (b) the force that must be applied at the tail when the distance between the center of pressure on the tail and the center of mass of the airplane is 15 ft.; (c) the mass that a similar engine of radius of gyration 14 in., would need to have in order that under similar conditions the pilot could produce a force on the tail amounting to 150 lb. wt.

Solution. (a) The torque is about a transverse axis and has a value

$$L = K_s w_s w_p$$

where

$$K_s \, [= mk^2] = \frac{325}{32.1} \left(\frac{14}{12} \right)^2 = 13.7 \text{ slug-ft.}^2$$

$$w_s = \frac{2 \pi \, 1300}{60} = 136 \text{ rad. per sec.}$$

and

$$w_p \left[= \frac{v}{r} \right] = \frac{220}{200} = 1.1 \text{ rad. per sec.}$$

Whence,

$$L \, [= K_s w_s w_p] = 13.7(136)1.1 = 2050 \text{ lb.-ft.}$$

(b) The force on the tail

$$F \left[= \frac{L}{X} \right] = \frac{2050 \text{ lb.-ft.}}{15 \text{ ft.}} = 136.5 \text{ lb. wt.}$$

60   MOTION OF A SPINNING BODY

(c) The weight of the larger engine is

$$mg = \frac{K_s g}{k^2} = \frac{\frac{Lg}{w_s w_p}}{k^2} = \frac{Lg}{k^2 w_s w_p}$$

$$= \frac{(150 \times 15 \text{ lb.-ft.})32.1 \text{ ft. per sec. per sec.}}{\left(\frac{14}{12}\text{ft.}\right)^2 136 \text{ rad. per sec. (1.1 rad. per sec.)}} = 355 \text{ lb. wt.}$$

Problem. The mass of a certain motorcycle is 350 lb., and the distance between wheel axles is 5 ft. Each wheel has a mass of 15 lb., diameter of 26 in., and radius of gyration of 12 in. There is a flywheel of 45 lb., and radius of gyration 5 in., with spin-axis in the direction of travel. The flywheel makes 3400 r.p.m. when the motorcycle is traveling 60 mi. per hr. The rider weighs 170 lb.

Neglecting the effect of any inclination of the machine from the vertical, and assuming that when the machine is at rest the wheels exert equal forces against the ground, find: (a) the gyroscopic torque tending to overturn the machine when making a curve of 300-ft. radius; (b) the vertical components of the reactions of the wheels against the ground at the same time.

Ans. (a) 22.3 lb.-ft.; (b) 255 and 265 lb. wt.

36. The Direction and Magnitude of the Torque Developed by a Spinning Body when Rotating About an Axis Perpendicular to the Spin-Axle. — Consider a gyro-wheel with axle vertical and

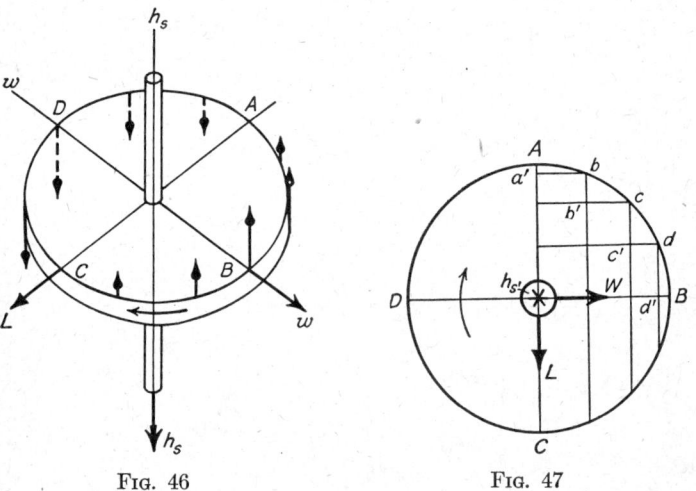

Fig. 46   Fig. 47

angular momentum h_s in the direction indicated in Fig. 46. Suppose that the gyro-axle is rotating with angular velocity w in the direction indicated. While the wheel is rotating, the particle at A is raised, the particle at C is lowered, while the particles along the axis of rotation of w are unaffected by the rotation about that axis. All particles of the wheel are moving about the spin-axis h_s. While a particle of the wheel moves from the point in space A to the point B, the component linear velocity of the particle changes in the direction of the spin-axis from a finite value to zero. Suppose that the points A, b, c, d, B, Fig. 47, are the positions of a particle after moving for equal intervals of time with constant speed about the spin-axle perpendicular to the plane of the diagram. If the plane of the wheel is at the same time rotating about an axis DB, the change in the magnitude of the linear velocity in the direction perpendicular to the plane of the wheel of a particle while moving from A to b is represented by Aa', and the changes while moving during succeeding equal intervals of time are represented by bb', cc', dd'. The rate of change of linear velocity increases as the particle moves from A to B. The inertia of the particle causes it to oppose this change of velocity by a force proportional to the rate of change of linear velocity of the particle. The forces exerted by the particle when at various points between A and B in opposition to the change of linear velocity normal to the plane of the wheel, are represented by arrows in Fig. 46.

While moving from B to C, the component of the velocity of the particle normal to the plane of the wheel increases from zero at B to a maximum at C. During the displacement from B to C, the inertia of the particle causes it to oppose this change of velocity at each point of its path by a force which is in the direction opposite to the velocity normal to the plane of the wheel. In Fig. 46, the forces at different points from B to C are represented by arrows. It is thus seen that each particle in the half of the wheel ABC tends to rise, thereby producing a rotation of the wheel about the torque-axis L. In a similar manner it can be shown that the half of the wheel CDA tends to rotate downward about the same axis.

Therefore, *when the axle of a spinning body is rotating about an axis perpendicular to the spin-axis, a torque is developed on the body about an axis perpendicular to the plane of the spin-axis and the axis of rotation. The direction of this internal torque is opposite that of the external torque which would need to be applied about the same axis in order to produce a precession of the gyro-axle in the direction in which the gyro-axis is rotating.*

The reaction of this torque acts in the opposite direction about the same axis on the agent that causes the rotation of the spin-axle.

When a torque is applied to an unspinning gyro about an axis perpendicular to the spin-axle, the spin-axle tilts in the direction of the torque. If the gyro is spinning and is unconstrained, the spin-axle precesses. This precession develops an internal torque that opposes the tilt which would occur if there were no precession.

About 1875 Sir Henry Bessemer organized a company to build a steamer to carry passengers across the English Channel without the discomfort of sea-sickness. Within the steamer was a large cabin weighing 180 tons mounted so as to be capable of oscillation about the fore-and-aft axis of the ship. A large wheel was mounted so as to spin about a fixed axis perpendicular to the floor of the cabin. The inventor assumed that the spin-axle would maintain its vertical position and thereby maintain the floor of the cabin nearly horizontal however the ship might roll. The device failed because the wheel was mounted so that it could not precess. Later, Schlick attained some success in preventing excessive roll of a ship by means of a large wheel spinning about an upright axis and capable of precessing about an athwartship axis (Art. 90).

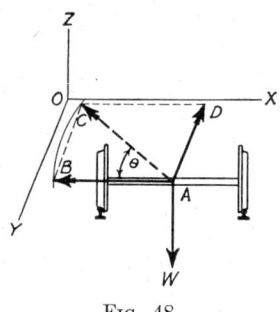

Fig. 48

Consider a pair of car-wheels, with the connecting axle, moving around a horizontal curve. Let the angular momentum h_s of the system with respect to the axle at any instant be represented by the line AB, Fig. 48, and the equal angular momentum at a short time Δt, afterward by AC. If the angular velocity about a vertical axis due to the motion around the curve be w, then the angle θ through which the axle turns in time Δt is

$$\theta = w\,\Delta t$$

The change in the angular momentum of the body, represented by the line AD, has the magnitude,

$$h_s\theta = h_s w\,\Delta t$$

and the time-rate of change of angular momentum, or the gyroscopic torque acting on the spinning body about the axis AD, in the clockwise direction, has the magnitude

$$L[=h_s w] = K_s w_s w \qquad (58)$$

The reaction of this torque, that is the torque developed by a spinning body of angular momentum h_s, when turned with angular

MOTION OF A SPINNING BODY

velocity w about an axis perpendicular to the axes of h_s and w, acts in the opposite direction with the same magnitude on the body that causes the change in the direction of the spin-axle:

$$L'[= h_s w] = K_s w_s w \qquad (59)$$

In the case of a car going around a curve, this reaction tends to rotate the car away from the center of the curve.

If the spin-axle is not perpendicular to the axis about which the spin-axle is turning, the torque is perpendicular to the axis of turning and to the component of the spin-velocity perpendicular to the axis of turning.

If w_s and w are of the same sign, the torque will be positive; if they are of opposite signs, the torque will be negative.

Problem. The shaft with the propeller of a certain torpedo boat has a moment of inertia of 2000 lb.-ft.^{2} with respect to the axis of the shaft. When the shaft is rotating at an angular speed of 300 r.p.m., the torpedo boat makes a half turn in 50 sec. Find the direction and magnitude of the torque developed on the bearings of the shaft.

Problem. A certain locomotive has four pairs of drive wheels, each pair with the connecting axle having a mass of 4000 lb., a diameter of 64 in., and a radius of gyration of 20 in. Find the value of the gyroscopic couple tending to lift the wheels on one side of the locomotive when the locomotive is moving with a speed of 60 mi. per hr. around a curve of 1000-ft. radius. Neglect any inclination of the axles to the horizontal.

Solution. From (58), the gyroscopic couple

$$L[= K_s w_s w] = 4\ mk^2 \frac{v}{r}\frac{v}{R} = \frac{4\ mk^2 v^2}{rR}$$

where m is the mass of one pair of wheels with the connecting shaft, k is the radius of gyration of the rotating system, v is the speed of the train, R is the radius of the curve, and r is the radius of each drive wheel.

$$L = \left(\frac{4(4000)}{32.1}\text{ slugs}\right)\left(\frac{20}{12}\text{ ft.}\right)^2 \left(\frac{60 \times 5280}{3600}\text{ ft. per sec.}\right)^2 \left(\frac{12}{32\text{ ft.}}\frac{1}{1000\text{ ft.}}\right)$$
$$= 4020\text{ lb.-ft.}$$

Problem. A steamship is propelled by three steam turbines rotating in the same direction about a fore-and-aft axis at the rate of 200 r.p.m. The moments of inertia of the three rotors together with the connected shafts and propellers are 1400, 700, and 1400 ton-foot^2 units, respectively. Find the magnitude of the gyroscopic couple developed when the ship is pitching with a maximum angular speed of 0.1 radian per second.

Solution. From (58), the torque $L = K_s w_s w$. Since one revolution equals 2π radians, the angular speed of the shafts is

$$w_s = \frac{200\ (2\ \pi)}{60}\text{ radians per sec.}$$

Whence, the torque developed by the pitching is

$$L[= K_s w_s w] = 2000 \left(\frac{1400 + 700 + 1400}{32.1} \text{ slug-ft.}^2\right) \left(\frac{400\,\pi}{60} \frac{\text{rad.}}{\text{sec.}}\right) \left(0.1 \frac{\text{rad.}}{\text{sec.}}\right)$$
$$= 2000(228) \text{ lb.-ft.} = 228 \text{ ton-ft.}$$

Problem. A certain passenger locomotive has a mass $m = 293{,}000$ lb.; distance of the center of mass from the plane of the rails $b = 7$ ft.; each of the four pilot wheels has a radius $r_1 = 20$ in.; each of the six drivers has a radius $r_2 = 40$ in.; each of the two trailer wheels has a radius $r_3 = 26$ in.; each pair of pilot wheels with the connecting axle has a mass $m_1 = 2100$ lb.; each pair of drivers with the connecting axle has a mass $m_2 = 8000$ lb.; the two trailers with the connecting axle have a mass $m_3 = 2700$ lb.; the radius of gyration of each pair of wheels with the connecting axle is 0.7 of the radius of the wheels. The distance between the centers of the rails, $i = 56.5$ in. $= 4.71$ ft.

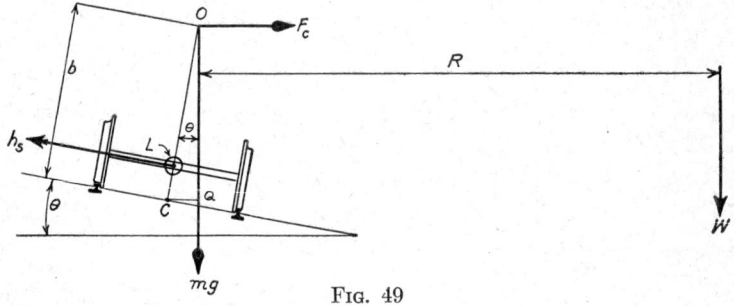

Fig. 49

Assuming that the locomotive is moving around a curve of radius $R = 2000$ ft., superelevated at an angle $\theta = 4°$, with a speed of 60 mi. per hr., find: (a) the centripetal torque, with respect to an axis parallel to the rails and midway between them, which tends to overturn the locomotive; (b) the gyroscopic torque, with respect to the same axis, due to the rotating wheels, which tends to overturn the locomotive; (c) the gravitational torque, with respect to the same axis, that opposes overturning. (d) With the locomotive running at the same speed along a straight level roadway, find the direction and magnitude of the gyroscopic torque developed when the wheels on one side pass over a high section of rail 1000 ft. long with a maximum elevation 0.1 ft. high at the middle, the other rail being horizontal.

Solution. Figure 49 represents a locomotive moving away from the observer and making a right-hand curve.

(a) In order that the locomotive may move around the curve, there must be a force acting at the center of mass directed horizontally toward the center of the curve of the magnitude

$$F_c = \frac{mv^2}{R}$$

Consequently, there must be a centripetal torque having a magnitude with respect to an axis perpendicular to the plane of the diagram through C, Fig. 49,

$$L_1[= F_c(OQ)] = \frac{mv^2}{R} b \cos \theta \tag{60}$$

If this torque be not applied, the locomotive tends to overturn, rotating away from the center of the curve about a horizontal axis perpendicular to the plane of the diagram. Substituting in (60) the data of the problem, we find the value of the centripetal torque tending to overturn the locomotive to be

$$L_1 = \frac{293{,}000}{32.1} \text{ slugs } \frac{(88 \text{ ft. per sec.})^2}{2000 \text{ ft.}} 7 \text{ ft. } (0.998) = 246{,}600 \text{ lb.-ft.}$$

(b) In order that a rotating axle may be turned about a vertical axis, it must be acted upon by a gyroscopic torque toward the center of the curve, about a horizontal axis perpendicular to the axle. The reaction of this torque acts upon the locomotive in the opposite direction (Art. 36).

The value of this torque acting on the locomotive, due to all six pairs of wheels, is

$$L_2 = (2 h_s' + 3 h_s'' + h_s''') w \cos \theta \qquad (61)$$

where h_s' represents the angular momentum of each of the pairs of pilot wheels with connecting axle, and h_s'' and h_s''' represent the angular momenta of the drivers and trailers, respectively.

From the data of the problem:

$$h_s'\left[= K_s' w_s' = m_s' k_1^2 \left(\frac{v}{r'}\right)\right] = \frac{2100}{32.1} \text{ slugs} \left(\frac{0.7(20) \text{ ft.}}{12}\right)^2 \frac{88 \text{ ft. per sec.}}{\frac{20}{12} \text{ ft.}} = 4701$$

Similarly, we find that

$$h_s'' = 35858 \quad \text{and} \quad h_s''' = 7858$$

$$w \cos \theta \left[= \frac{v}{R} \cos \theta \right] = \frac{88 \text{ ft. per sec. } 0.998}{2000 \text{ ft.}} = 0.044 \text{ radian per sec.}$$

Hence the magnitude of the gyroscopic torque is

$$L = [2(4701) + 3(35858) + 7858]0.044 = 5493 \text{ lb.-ft.}$$

(c) The gravitational torque about a horizontal axis parallel to the rails is clockwise and has the magnitude

$$L_3 = mgb \sin \theta$$
$$= 293000 \text{ lb. wt. } (7 \text{ ft.})0.07 = 143{,}570 \text{ lb.-ft.} \qquad (62)$$

(d) When one side of the locomotive ascends a high spot, the axles rotate about a horizontal axis parallel to the straight rail with angular velocity w', of the magnitude, Fig. 50:

$$w' = \frac{\phi}{t} = \frac{e}{jt} \text{ radians per sec.}$$

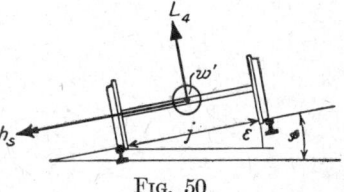

Fig. 50

The linear velocity of the train while ascending the elevation, Fig. 51,

$$v = \frac{l}{t} \text{ ft. per sec.}$$

so that

$$w' = \frac{ev}{jl} \text{ radians per sec.}$$

Consequently, when the wheel on one end of an axle ascends an elevated spot, there is developed a gyroscopic torque about an axis perpendicular to the plane of the axis of w' and of the spin-velocity w_s. The direction of this torque about its axis is opposite that of the torque which would need to be applied about the same axis in order to produce a precession of the axle in the direction of w'. This torque urges the ascending wheel in advance of the wheel on the level rail. The magnitude of this gyroscopic torque due to a single pair of wheels with connecting axle is

$$L_4 \, [= h_s w'] = h_s \frac{ev}{jl} \qquad (63)$$

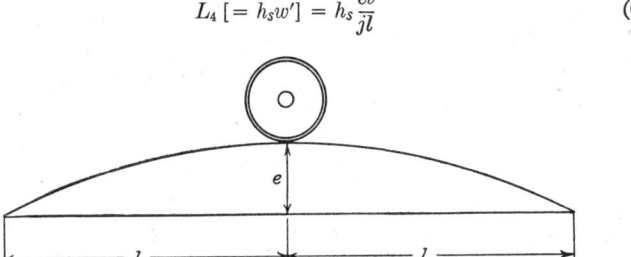

Fig. 51

The torque developed by the six wheels on one side of the locomotive ascending the bump is

$$\Sigma L_4 = [2\, h_s' + 3\, h_s'' + h_s''']\frac{ev}{jl}$$
$$= [2(4701) + 3(35858) + 7858]\frac{0.1(88)}{4.71(500)} = 462 \text{ lb.-ft.}$$

On descending the bump an equal torque is produced in the opposite direction about the same axis. These torques tend to spread the rails.

37. The Period of Precession. — The period of precession, that is the time T required by the spin-axle to make one complete revolution with constant angular velocity w_p about the axis of precession, is given by the expression

$$\theta [= T w_p] = 2\,\pi \text{ radians}$$

When the spin-axle and the precessional axis are perpendicular to one another, the maintenance of the precessional velocity requires a torque about an axis perpendicular to the spin-axle and the precession axis of the value, (55), $L = K_s w_s w_p$. Consequently,

$$T\left[= \frac{2\,\pi}{w_p}\right] = \frac{2\,\pi K_s w_s}{L} = \frac{2\,\pi h_s}{L} \qquad (64)$$

38. The Kinetic Energy of a Precessing Body. — The kinetic energy of a gyro-wheel of moment of inertia K_s spinning with an angular velocity w_s is $\frac{1}{2}\, K_s w_s^2$, (11). So long as w_s remains constant, the kinetic energy due to spinning is constant, whether there is precession or not. If a mass m be applied to one end of the spin-axle of the gyroscope, the spin-axle will dip through an angle ϕ, and a torque L_t will be developed about the torque-axis. The spin-axle will precess about an axis perpendicular to the spin-axle and the torque-axis.

A torque imparts kinetic energy of rotation to a body only when the torque produces angular velocity about the torque-axis. Consequently, the torque L_t imparts kinetic energy to the spinning body only in producing the dip ϕ. Hence, the kinetic energy of precession equals the work done by the torque L_t in producing the dip ϕ and has a value equal to $L_t\phi$. Therefore, the total kinetic energy of the precessing body is

$$W_{ro} = \tfrac{1}{2} K_s w_s^2 + L_t \phi \qquad (65)$$

The spin-velocity w_s of a gyro-wheel is not altered by the application of a torque that produces a precession of the gyro-axle.

In uniform precessional motion, torque produces angular momentum about the axis of precession but does no work about that axis. This is analogous to the case of uniform circular motion in which centripetal force produces linear momentum but does no work in the direction of the linear momentum.

39. Nutation. *Experiment.* — Figure 52 represents a gyroscope mounted as used by Fessel. By setting the counterpoise at different places on the rod, torques about a horizontal axis can be produced of different magnitudes in either the clockwise or the counter-clockwise direction. By means of the counterpoise, balance the gyroscope so that the rod remains horizontal. Set the gyro-wheel spinning. Observe the result.

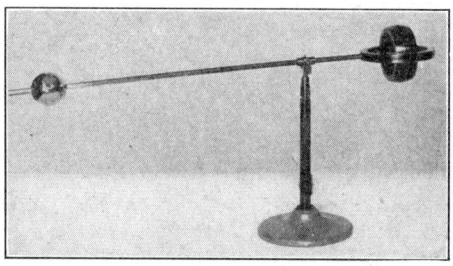

Fig. 52

Stop the spinning. Move the counterpoise till it overbalances the gyro. Set the gyro spinning. Hold the rod horizontal and then release. Observe that the end dips and rises and at the same time precesses with ununiform velocity. The nodding of the spin-axle is called *nutation*.

Stop the spinning. Move the counterpoise till it is overbalanced by the gyro. Set the gyro spinning. Hold the rod horizontal and then release. Observe the result.

Any gyroscope exhibits nutation of the spin-axle on the sudden application of a torque, but the long arm of Fessel's mounting renders the nodding especially conspicuous. The amplitude of the nutational dip depends upon the amplitude and suddenness of development of the gravity torque. Nutation would continue indefinitely if there were no opposition to the dipping due to air or bearing resistance, as for example in the case of the nutation of the earth's axis. In the case of most gyroscopes the nutation is quickly damped to zero.

Suppose that the gyro is spinning at constant rate in the clockwise direction as seen when viewed from the counterpoise toward the gyro. In addition, suppose that the position of the counterpoise is such that there is a torque L in the clockwise direction about a horizontal axis as indicated in Fig. 53. Note that there is zero external torque about the vertical axis.

At the instant the gyro-frame is released, there is zero precession and the heavy end tilts downward below the horizontal. This tilting gives a vertical component Δh_s of the angular momentum h_s, directed downward, Fig. 53. There must be zero torque about the vertical axis. Another angular momentum must be generated about the vertical axis at the same rate and in the direction opposite to the vertical component of h_s. This is produced by the development of angular velocity (and angular momentum) in the direction indicated by w_p. This precessional velocity is greater than that which would have been produced if the spin-axle had not been tilted so much.

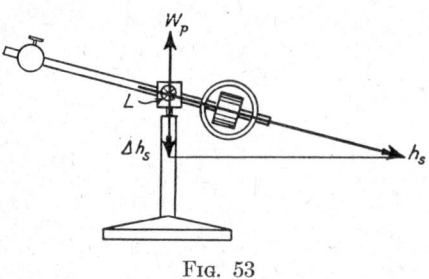

Fig. 53

The gyroscope has greater kinetic energy when precessing than when not precessing, though the total energy is constant. In order that the total energy may remain constant, the accelerated precessional velocity is accompanied by a rise of the heavy end of the gyroscope.

The inertia of the system carries the heavy end above the equilibrium position. As the potential energy is increased by the elevation of the heavy end of the gyroscope, the kinetic energy of rotation must decrease, that is, the precessional velocity must diminish.

MOTION OF A SPINNING BODY 69

When the heavy end dips, the precessional velocity increases. When the heavy end rises, the precessional velocity decreases. After a few oscillations, the spin-axle will remain at a nearly constant inclination to the vertical and the velocity of precession will remain nearly uniform. The inclination of the spin-axle from the vertical, at any instant, is called the angle of nutation.

Question. The counterpoise of Fig. 52 may be replaced by a little bucket of sand provided with a hole in the bottom through which the sand may slowly escape. Suppose that this bucket is placed near the end of the rod so as to considerably overbalance the gyroscope, the gyro is set spinning, the rod is held horizontal and then released. As the sand escapes, the counterpoise will, after a time, be overcome by the weight of the gyro. Describe the sequence of changes in precession and dip of the gyro-axle.

40. The Effect of Hurrying and of Retarding the Precession of a Spinning Body. *Experiment.* — Attach a small mass to the inner frame of the gyroscope at the point x, Fig. 35, thereby producing a torque L as indicated. Set the gyro-wheel spinning in the direction indicated. Observe that the weighted side of the inner frame dips slightly below the center of the gyro-wheel and that the gyro-axle precesses with an angular velocity in the direction represented by w_p. Push horizontally against the second gyro-frame with the rubber tip of a lead pencil so that the spin-axle is moved in the direction of its precession. Observe that the weighted side of the inner gyro-frame rises. Now push on the second frame so that the spin-axle is moved in the direction opposite its precession. Observe that the weighted side of the inner frame sinks.

In each case the spin-axle moved so as to be more nearly parallel to the axis of the new torque produced by the pushing on the second frame, and with the direction of spin in the direction of the torque. This is in accord with the Second Law of Gyrodynamics (Art. 34). When an added torque causes an increase in the precessional velocity, the angular momentum produced by the added torque about the precession axis is represented by the line Op, Fig. 54. This angular momentum, compounded with the angular momentum OS of the gyro-wheel about the spin-axis, gives a resultant angular momentum represented by OR. Thus, it appears that the spin-axis tends to precess upward through an angle θ about an axis through O perpendicular to the plane of OP and OS. By a similar analysis it can be

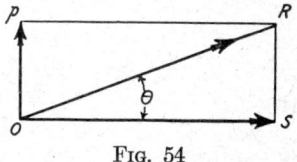

FIG. 54

shown that an added torque that retards the original precession will develop another precession that causes the spin-axis to dip (Fig. 55).

The precession of the gyro brings into play a torque which neutralizes the torque due to the overweight (Art. 36), and, so long as the precessional velocity remains constant, the overweight will not descend. If the precessional velocity be diminished, by pushing on the frame for example, the torque due to precession will be less than sufficient to counteract the torque due to the overweight, and the overweight will descend. If the precessional velocity be increased, the torque due to precession will be more than sufficient to counteract the torque due to the overweight, and the overweight will rise.

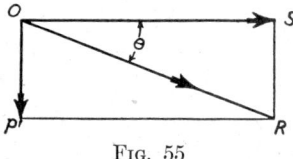

Fig. 55

Air resistance and friction of bearings always produce a torque about the precession axis in the direction opposite the precession. It is for this reason that the spin-axle of a balanced gyro tends to dip.

(a) When the precessional speed of the axle of a spinning gyro is increased, the gyro is acted upon by an internal torque in opposition to the torque that produces the precession.

(b) When the precessional speed of the axle of a spinning gyro is decreased, the gyro is acted upon by an internal torque in the same direction as the torque that produces the precession.

(c) When an external torque is applied to the axle of a spinning gyro in the direction of the precession, an internal torque is developed which acts upon the gyro in opposition to the torque that produces the precession.

(d) When an external torque is applied to the axle of a spinning gyro in the direction opposite to that of the precession, an internal torque is developed which acts upon the gyro in the same direction as the torque that produces the precession.

While a spinning top is held on the palm of the hand, if the hand be moved in a horizontal circle in the direction of the precession, the top will tilt toward the vertical. " Hurry the precession, and the top rises; retard the precession and the top falls." This applies to a top spinning at an ordinary rate. If the spin-velocity be so great that the fast precessional velocity occurs (Art. 35), then if the precession be hurried, the spin-axle will tilt away from the vertical.

As soon as a moving bicycle leans to the right, the rider turns the machine about a vertical axis to the right. The spin-axes of the wheels tend to set themselves parallel to this vertical torque-axis, thereby tilting the machine upward. Usually the bicycle will tilt beyond the vertical to the left; the same series of operations will be repeated in the opposite direction; the machine will oscillate back and forth about the position of equilibrium, and the track on the ground will be serpentine. This gyroscopic righting effect is less than the righting effect due to the tendency of the center of mass to continue in the direction in which it was already going.

The above law (*a*) is the basis of devices for preventing rolling of a ship in a seaway (Arts. 92–97) and for preventing the overturning of a car when running on a single rail below the car (Chap. VI). Law (*b*) is the basis of a device to cause a ship to roll when it is desired to get off a sand bar, or to break a passage through a frozen harbor (Art. 96).

41. Motion of the Spin-Axle Relative to the Earth. — According to the First Law of Gyrodynamics (Art. 34), the spin-axle of a gyro mounted so that the intersection of the three axes of rotation coincides with the center of mass will remain fixed in space, however the frame may be displaced, until acted upon by a torque. If the spin-axle is parallel to the axis of the earth, it will continue to be parallel while the earth rotates. In general, however, it will not remain in fixed relation to the earth. Consider a gyro with spin-axle horizontal. If it be situated at either geographic pole, it will rotate relative to the earth, making one revolution in a horizontal plane in one day. If it be situated at the equator, in the plane of the meridian, it will remain horizontal and in the meridian line. If it be situated at the equator and in the plane of the equator, it will remain in the plane of the equator but will make one revolution in one day about an axis perpendicular to the plane of the equator. At any latitude between 0° and 90°, the spin-axle will rotate relative to the earth about a vertical axis in the counter-clockwise direction as seen from above, and also will dip about a horizontal axis.

Either of these two angular motions of the gyro-axle can be prevented by the action of a suitable torque. If the torque causes a precession just equal to the vertical component of the angular velocity of the earth, then the spin-axle will turn with the earth and will maintain its position with respect to the earth. Thus suppose that at a certain latitude, the angular velocity of the gyro-axle relative to the earth, about a vertical axis, is w radians per second. Then motion of the gyro-axle in azimuth can be prevented by the application of a torque about the east-west axis that will produce a precessional velocity $w_p = w$, given by the Second Law of Gyrodynamics:

$$L = w K_s w_s$$

where L is the torque expressed in pound-feet, K_s is the moment of inertia of the gyro relative to the spin-axle expressed in slug-feet^2, and w_s is the spin-velocity expressed in radians per second.

If angular velocity of the spin-axle relative to the earth expressed in degrees per minute of time be represented by w', moment of inertia of the gyro relative to the spin-axle expressed in pound-feet^2 be represented by K_s', and spin-velocity expressed in revolutions per minute be represented by n, then the above equation may be written

$$L[= w K_s w_s] = \frac{w'}{(57.3)60} \frac{K_s'}{32.1} \frac{2\pi n}{60} = 0.00000095 \, w' K_s' n \quad (66)$$

The vertical component, relative to the earth, of the angular velocity of the spin-axle of a freely suspended gyro in latitude λ is

$$w = w_e \sin \lambda \quad (68)$$

where w_e represents the angular velocity of the earth about the geographic axis.

Since the angular velocity of the earth is $[360° \div 24 \times 60] = 0.25°$ per minute of time,

$$w' = 0.25 \sin \lambda \text{ degrees per minute of time.} \quad (69)$$

42. The Sperry Directional Gyro. — The indications of a magnetic compass on an airplane can be relied upon only when the airplane is steady and moving in a straight course with constant speed. In taking a compass reading, the common practice is to steady the airplane in a straight course when near the desired compass heading and wait till the oscillations of the needle have become sufficiently small. To maintain a straight course either the ground must be visible or some gyroscopic turn-indicator must be available.

The various uncertainties in the indications of the magnetic compass have caused it to be largely displaced by the gyro-compass for navigating marine vessels, but the considerable weight and cost of the gyro-compass have prevented its adoption on airplanes. Many of the advantages of the gyro-compass for navigational purposes can be obtained by a simple gyroscopic device which is ckecked and reset at intervals against the indications of a magnetic compass.

The Sperry Directional Gyro consists of a freely suspended gyro G, Fig. 56, capable of spinning at a speed of some 12,000 revolutions per minute about an axis which normally is horizontal. At this speed of rotation, the spin-axle will maintain its direction in space within three degrees for a quarter hour. The spin of the gyro is maintained by air currents from two nozzles directed against

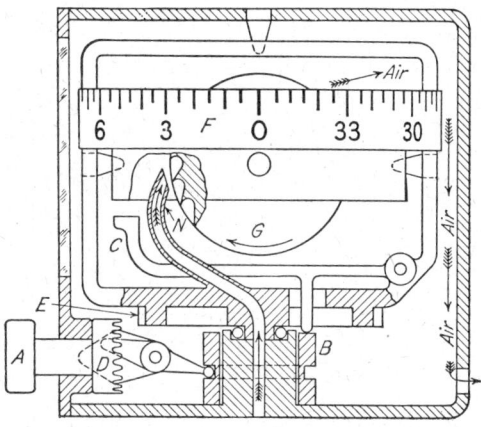

Fig. 56

buckets cut in the edge of the gyro. The air flow is produced by a venturi as represented in Fig. 69, page 85. The path of air through the instrument is indicated by the feathered arrows.

In using the directional gyro as a navigational instrument, one method of procedure is first to put the airplane on the desired course by reference to the compass, then push in the setting knob A, thereby lifting the ring B and the lever C which, in turn, brings the spin-axle horizontal and locks it in that position. Now, the indicating card F of the instrument may be brought to the zero position by twisting the setting knob which, in turn, rotates the gear D, the azimuthal gear E and the attached indicating card. Pulling out the setting knob frees the gyro from all constraint. Then, the airplane can be kept in the desired course by steering so that the card of the directional gyro continues to indicate zero.

Problem. A gyro of moment of inertia with respect to the spin-axis of 128 lb.-ft.² is spinning at latitude 45° N. with an angular speed of 2000 r.p.m. The gyro is mounted so that the center of gravity of the wheel and that of each of the supporting gimbal rings coincides with the intersection of the axes of the knife-edges of the two rings. Find the mass and the position that an added body must have so that when placed 4 in. from the center of mass of the gyroscope it will cause the spin-axle to remain in the meridian plane.

43. An Instrument for Measuring the Crookedness of a Well Casing.

— In drilling or in boring a deep well it is very difficult to maintain the bore hole vertical or straight. Crooked holes may prevent the rotation of the drill or may cause the drill to twist off. The bottom of a well a couple of thousand feet deep may be more than a hundred feet to one side of the vertical through the top of the hole. It may be on the property of a proprietor other than the one who owns the land on which the upper end of the well is situated. Until recently no method has been known to measure the direction and amount of the bends of a deep well bore.

By means of the Surwel gyroscopic clinograph, a permanent record of the position, direction, length and amount of any bends in a well casing can now be obtained. The apparatus is enclosed within a cylindrical case several feet in length and five and one-half inches outside diameter. This case contains a universal spirit level, a universally mounted gyro with horizontal spin-axle, a watch, a movie camera, miniature incandescent lamps, and batteries of dry cells to spin the gyro and operate the lamps and movie camera.

The upper spherical surface of the spirit level carries a series of concentric circles spaced so as to indicate the inclination of the axis of the case from the vertical in any direction. The bubble cell is provided with an expansion coil which keeps the bubble the same size whatever the temperature of the apparatus.

Any turning of the instrument about its geometrical axis while being lowered into the well is indicated by the gyro which spins about a horizontal axis at a speed of about 11,000 revolutions per minute. The natural turning of the spin-axle relative to the earth, due to the rotation of the earth, is neutralized by a precession of equal speed in the opposite direction produced by the weight of a small mass attached to the north bearing of the spin-axle (Art. 41). The vertical pivot of the gyro-casing carries a pointer which indicates the direction, in azimuth, of the spin-axle. Besides the two hands of the watch there is a pointer that marks the time the apparatus started down the well.

The instrument is lowered into the well at a known constant rate. The movie camera takes a series of pictures at a regular interval which depends upon the speed with which the instrument is lowered. The first picture is taken as the instrument starts down the well. Each picture shows the time as well as the inclination of the instrument and the amount the instrument may have twisted about its axis from its original position, at the instant the picture was taken. From the views of the three scales shown in any one picture the distance of the apparatus from the top of the well at any instant can be determined, as well as the direction and magnitude of the inclination from the vertical of the bore hole at that instant.

44. The Weston Centrifugal Drier. — In laundries water is removed from wet clothes by a "centrifugal drier" consisting of a metal cylinder with perforated sides capable of rotating at high speed about a vertical axis. In most models, the perforated cylinder is rigidly fastened on the upper end of a vertical shaft. If the wet clothes are not packed uniformly about the axis of rotation the shaft will be subjected to considerable stresses when rotated at high speed.

In the Weston centrifugal machine used to separate molasses from sugar crystals, this difficulty is avoided by hanging the perforated cylinder from a universal joint on the lower end of a vertical shaft. In case the center of mass of the suspended system is not on the axis of rotation, it will move away from the vertical line through the point of support, Fig. 57. The suspended system now is acted upon by a gravitational torque in the direction represented by the symbol at L in the figure. When

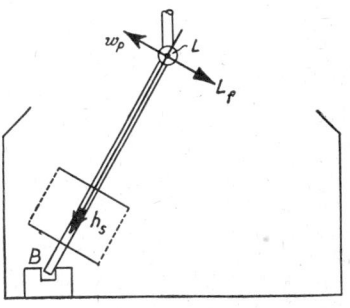

Fig. 57

the suspended system is rotating in the direction represented by h_s, it will precess in the direction represented by w_p.

A pin, in line with the axis of the perforated cylinder, fits loosely in a block of metal B that can slide around on the bottom of an outer cylinder. When the axis of the perforated cylinder is not vertical, this block is dragged around in a circular path, thereby developing a torque on the suspended system. At the instant represented in the figure, this frictional torque is represented by L_f. At all times the torque due to friction will be in the direction opposite to the precession w_p. It follows, from law (d), Art. 40, that the suspended system is acted upon by a torque in the same direction as the gravitational torque L. Consequently, the suspended system precesses so as to hang more nearly vertical.

45. The Effect on the Direction of the Spin-Axis of a Top Produced by Friction at the Peg. — Figure 58 represents the rounded peg of a top that is spinning about an axis which is inclined to the vertical and is in the vertical plane XZ. The center of mass of the top is at C.

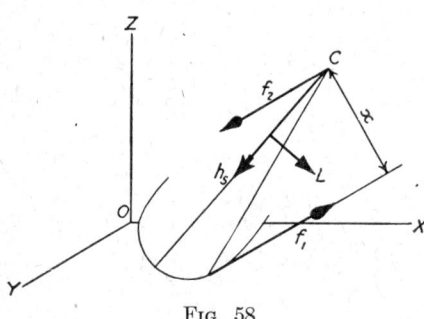

Fig. 58

Let h_s represent the angular momentum of the top relative to the spin-axis. The force of friction f_1 at the point of contact of the peg with the ground is shown parallel to the OY axis and is directed away from the reader. An equal parallel force f_2 acts in the opposite direction at the center of mass C. These two forces constitute a force couple having a lever arm x. The moment of this couple is about an axis in the vertical plane and is in the direction indicated by L.

Since the spin-axis tends to set itself parallel to and in the same direction as the torque-axis, it follows that h_s turns about an axis parallel to YO in the counter-clockwise direction as viewed by the reader. As h_s becomes more nearly vertical, x and therefore L decreases in value.

If the spin-axis were inclined to the left of the vertical, the direction of the torque due to friction would be in the direction opposite that represented in the figure, and the spin-axis would tend to become vertical as before.

Thus, on account of the friction between the ground and the rounded peg of a spinning top, the spin-axis of the top tends to become vertical. Owing to this effect, the upper end of the precessing axis of the top traces a converging spiral.

46. The Bonneau Airplane Inclinometer. — There is great need of an instrument that will furnish a line which will remain truly vertical when on a ship or airplane subject to large linear or angular accelerations. The Bonneau inclinometer* is a top having a spherical peg the center of which coincides with the center of gravity of the top, Figs. 59 and 60. The peg stands in a spherical cavity of less curvature. The top is free of gravitational torque and precessional motion. Centripetal and other lateral forces do not deflect the spin-axle unless they displace the pivot in its bearing. If such a displacement occurs, the rounded supporting peg accelerates any precessional velocity thereby causing the spin-axle to assume a practically vertical position (Art. 45). The top is maintained in rotation by two streams of air impinging on short blades cut in the periphery. The driving air currents are produced by the motion of the airplane through the air.

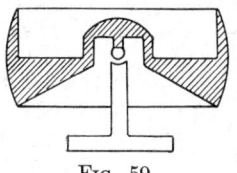

Fig. 59

In one model, there are two parallel circular stripes about the periphery of the top and another similar stripe about the cylin-

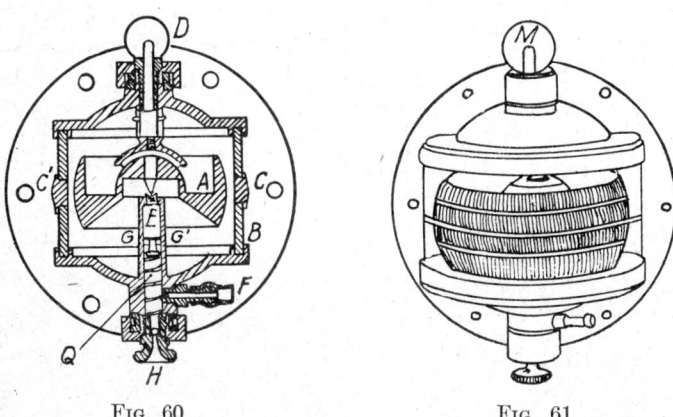

Fig. 60   Fig. 61

drical glass housing, Fig. 61. When the airplane is horizontal, the stripe on the housing is parallel to the two stripes on the top.

In another model, the indication is made by a light pointer fastened to the top coaxially with the spin-axis. On the spherical cover glass over the top there are circles and radial lines by means of which the observer can read the inclination of the housing to the vertical gyro-axle.

* U. S. Patent. Bonneau, No. 1435580, 1922.

78 MOTION OF A SPINNING BODY

47. The Sperry Airplane Horizon. — It is impossible for an aviator surrounded by dense clouds to determine without instruments whether the airplane is on even keel, is inclined to the horizontal, or even whether he and the airplane are upside down. The Sperry airplane horizon is an instrument designed to indicate the angle between the keel of the airplane and the horizon. It consists of a gyro spinning at about 12,000 r.p.m. about an axis which normally is vertical. The gyro casing is mounted in a gimbal ring having axes parallel to and perpendicular to the longitudinal axis of the airplane. The intersection of the gimbal axes is at a short distance above the center of mass of the gyro and casing.

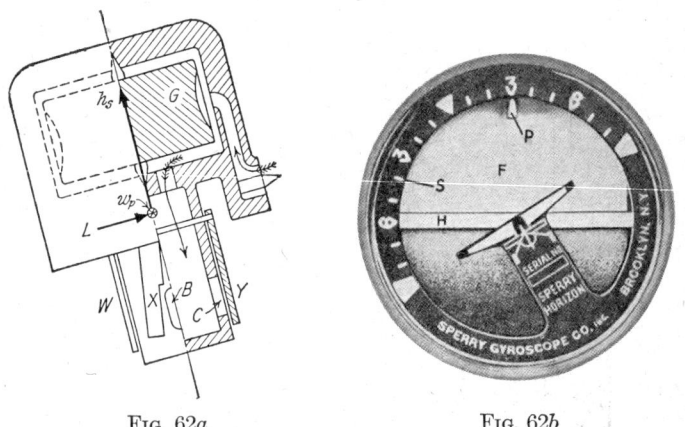

Fig. 62a   Fig. 62b

The rotation of the gyro is maintained by the impact of two jets of air directed tangentially against a row of buckets cut in the edge of the gyro. The air is set into motion by a venturi which draws air from the surroundings through ducts in one of the gimbal trunnions, the gimbal ring and the two nozzles. A venturi used in a similar manner is illustrated in Fig. 69.

The exhaust air escapes from four ports, spaced 90° apart, through the lower part of the gyro casing. Hanging pendulum-wise in front of each port is a vane or shutter which varies the effective area of the port behind it as the casing tilts. When the gyro is vertical, each port is covered to the same degree and the reactions on the casing due to air escaping from the four ports are balanced. When, however, the gyro-axle is not vertical, the ports are covered by the pendulous vanes in unequal degree and the

MOTION OF A SPINNING BODY

reactions due to the escaping air are not balanced. With the gyro tilted as in Fig. 62a, the port B is not covered by its pendulous vane X whereas the other three ports are covered by their respective vanes. The reaction of the escaping air urges the lower end of the gyro casing away from the reader, thereby producing a torque in the direction represented by the directed line L. This torque, when the gyro is spinning in the direction represented by the directed line h_s, results in a precession w_p which brings the spin-axle into the vertical position without oscillation.

The face of the Sperry airplane horizon has a vertical dial F, Fig. 62b, fastened rigidly to the gimbal ring. The upper part of the dial is colored blue to represent the sky and the lower part is colored grey to represent the earth. The dial is encircled by a mask which is fastened to the case of the instrument. The mask carries a divided circular scale and also a silhouette of an airplane as viewed from the tail.

When the airplane banks, that is, tilts about a fore-and-aft axis, the case of the instrument rotates relative to the gimbal ring and to the gyro spin-axis. The angle of bank is indicated by the position of the pointer P marked on the air-earth dial, relative to the circular scale S attached to the instrument case.

When the nose of the airplane is directed downward, the actual horizon appears to the pilot to rise, and when the airplane is climbing, the actual horizon appears to sink. A white bar H, in front of the dial of the instrument, fastened to a link which is pivotted to the gimbal and to the case of the instrument, rises and sinks relative to the center of the dial, exactly as the actual horizon appears to rise and sink relative to the nose of the airplane. Fig. 62b indicates a 30° left bank and level fore-and-aft axis of the airplane.

The horizon bar defines a horizontal plane. The angle between this plane and the keel of the airplane depends upon the loading of the airplane. For example, when the airplane is flying a level course with tail high, the Sperry airplane horizon indicates a slight dive, whereas when the airplane is loaded to such an extent that it rides tail low, the horizon bar indicates a slight climb.

When the airplane is making a turn, the pendulous shutters on the gyro casing would cause an unbalanced reaction on the casing even though the spin-axle were vertical. This results in a slight precession of the spin-axle and tilt of the horizon bar which, however, ceases as soon as the airplane resumes a straight course.

The position of the horizon bar used as normal is that when the airplane is in straight level flight with its present load. By the aid of standard turn indicator and rate-of-climb indicator, the pilot knows when the airplane is in straight level flight. Knowing the normal position of the horizon bar, the pilot can use the Sperry horizon to inform him of the altitude of his airplane during night or blind flying when the invisibility of the actual horizon would render flying difficult and often dangerous.

48. The Drift of a Projectile from a Rifled Gun. — If a projectile fired from a gun were uninfluenced by any force except its weight, the center of mass of the projectile would describe a parabola having a vertical axis (Ferry's General Physics, p. 68). If the bore of the gun were not rifled, there would be no tendency for the projectile to rotate about any axis. If the projectile were cylindrical, the axis would remain parallel to the axis of the gun throughout the flight of the projectile, as represented in Fig. 63.

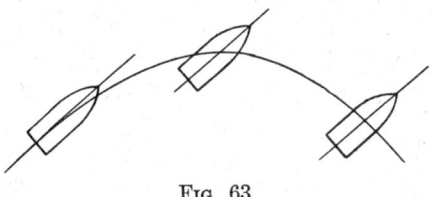

Fig. 63

In the actual case, however, the projectile moves through air and is acted upon by air resistance, which profoundly modifies the motion. Air resistance lifts the nose of the projectile. If the bore of the gun were smooth, the nose of the projectile might be lifted to such an extent that the projectile would strike broadside-on or might tumble. For this reason, the projectiles used with smooth-bore guns are spherical. By rifling the bore of the gun, the projectile is given an angular velocity of spin about the long axis of as much as 3000 revolutions per second for infantry arms. The large projectiles of the coast artillery spin much more slowly, in some cases as slowly as 100 revolutions per second. The projectiles from American guns rotate in the clockwise direction as viewed from the gun. Owing to the "rigidity of the spin-axis," a spinning long projectile has a much greater range than if it were not spinning, but there is a lateral deflection. With projectiles of the same model, this lateral drift is to the right when the direction of spin is clockwise and to the left when the spin is counterclockwise. The amount of drift depends upon the shape and upon the angular momentum of the projectile about the spin-axis. The drift amounts to as much as one yard in a 1000-yard range and eleven yards in a 3000-yard range. Drifts as great as 2000 feet have been observed in a 20-mile range. Drift is caused by a combination of the direct effects of air pressure and air friction, and by gyroscopic effects produced by torques developed by air resistance and air friction. There is no complete explanation of the observed facts of drift of projectiles, but it is probable that the following actions are important causes.

Due to the rotation of the earth, a body moving with linear velocity over the surface of the earth will be deflected, relative to an observer on the earth, out of its original path. The vertical component of the linear velocity of a projectile is deflected westward when the projectile is ascending and is deflected eastward when descending. A body thrown vertically upward will fall west of the place of projection. The horizontal component of the linear velocity of a projectile is deflected to the right in northern latitudes and to the left in southern latitudes. These deflections are not due to forces acting on the projectile but are due solely to the rotation of the body to which the linear velocity of the projectile is referred.

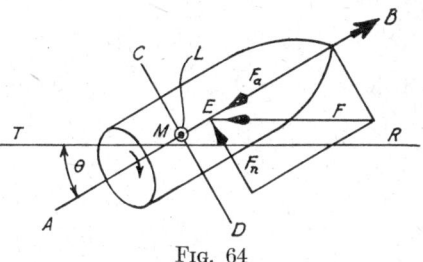

FIG. 64

Consider a projectile spinning in the clockwise direction about the axis of figure. There is a force due to air friction opposing the spin. When the axis of the projectile is inclined to the trajectory, this air friction is greatest on the under side of the nose and the projectile "rolls off" toward the right. Since the friction is greater on the nose than on the rear end, the front end of the projectile rolls to the right more rapidly than the rear end. The wind now striking the left side of the projectile causes a drift to the right.

Again, suppose that the axis of figure of the projectile is inclined to the trajectory TR, at an angle θ, Fig. 64. The resultant air resistance is a force F parallel to the trajectory and applied at a point called the center of effort. Usually the shape of a projectile is so designed that the center of effort E is on the axis of figure and in advance of the center of mass M. Resolving F into two components, one parallel and one normal to the axis of the figure, we have $F_p = F \cos \theta$ and $F_n = F \sin \theta$. The former component retards the motion of the projectile. The latter has a moment L about the center of mass, $L = xF_n = xF \sin \theta$, where x represents the distance ME. This torque $xF \sin \theta$ acts about an axis through M perpendicular to the plane of the axis of figure and the trajectory, tending to raise the nose of the projectile. The angular momentum of the projectile about the axis of figure is so great, however, that this torque produces negligible deflection about the torque-axis but does produce a precessional velocity about an axis perpendicular to the torque-axis and the spin-axis. In Fig. 65, the angular momentum h_s, the torque L, and the precessional velocity w_p are represented by directed lines. Since the spin-axis tends to set itself parallel to the torque-axis, with the direction of spin in the direction of the torque, it follows that, owing to air pressure on the under side of the nose of the projectile, the nose is deflected to the right along a line perpendicular to the plane containing the trajectory and the spin-axis.

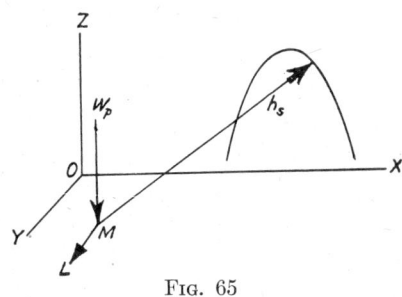

Fig. 65

Again, the friction of air against the lower side of the nose of the spinning projectile produces a torque opposing spinning, about an axis CD, Fig. 64, nearly parallel to the plane containing the trajectory and the spin-axis. On account of this torque, the spin-axis AB precesses about an axis perpendicular to the plane of the spin-axis and torque-axis, tending to become parallel with and in the same direction as the torque-axis CD. Thus, the nose of the projectile is lowered and the spin-axis is made more nearly tangent to the trajectory.

The result of the drift and the drop is a rotation of the axis of the projectile in the clockwise direction about the trajectory. The rotation of the axis of the projectile would be counter-clockwise if either the direction of spin were counter-clockwise or the center of effort were behind the center of mass.*

49. The Effect of Revolving a Non-Pendulous Gyroscope with Two Degrees of Rotational Freedom about the Axis of the Suppressed Rotational Freedom. — A gyroscope in which the center of mass of the gyro and supporting frame is at the intersection of the axes of rotation is said to be non-pendulous. If the center of mass is below the intersection of the axes of rotation, the gyroscope

---

* For a fuller discussion see Crantz, Lehrbuch der Ballistik, 1925; Hermann, Exterior Ballistics; Moulton, New Methods in Exterior Ballistics, 1926.

is said to be pendulous; if it is above, the gyroscope is said to be antipendulous or top-heavy.

Experiment. — Clamp the middle ring of the gyroscope, Fig. 34, so that rotation of this ring about the vertical axis is prevented. As the gyro now can rotate about two axes perpendicular to one another but not about a third perpendicular to the plane of the other two, the gyro is said to have two degrees of rotational freedom. Set the gyro spinning and place the gyroscope on a rotatable platform. Rotate the platform in the clockwise direction about the vertical axis. Observe that the spin-axle of the gyro sets itself parallel to the vertical axis with the direction of spin in the clockwise direction. Now rotate the platform in the counter-clockwise direction. Observe that the gyro turns over so that the spin-axle is again parallel to the vertical axis, and the direction of spin is in the counter-clockwise direction.

FIG. 66

When the gyro-axle is revolved, a torque acts upon the gyro relative to the axis about which rotation is suppressed. The spin-axis tends to set itself parallel to this torque axis, and with the direction of spin in the direction of the torque, according to the Second Law of Gyrodynamics (Art. 34).

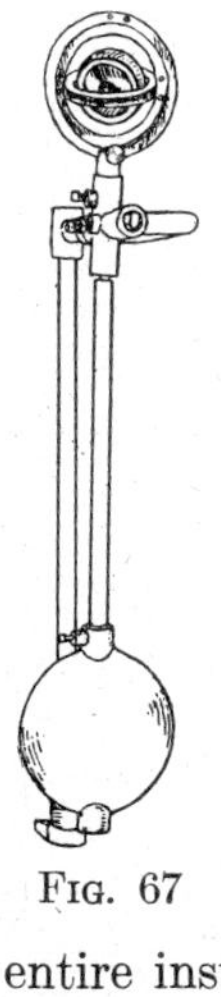

Experiment. — The gyro of Fig. 66, has but two degrees of rotational freedom. The arrow is fixed rigidly to the frame supporting the wheel so as always to be in the same direction as the gyro-axle. If a person holding the handle of the instrument horizontal rotates himself about one heel, the gyro-axle with the attached arrow will set itself vertical. When he turns about the heel in the opposite direction, the wheel with the attached arrow will turn a somersault. In each case the axle becomes parallel to the axis of rotation of the entire instrument and with the spin of the wheel in the same direction as the rotation of the instrument.

FIG. 67

Experiment. — Figure 67 represents a knife-edge pendulum to one end of which is attached a gyroscope in a frame that cannot rotate about an axis parallel to the knife-edge. Set the pendulum swinging and observe that the gyro-axle sets itself parallel to the knife-edge and with the direction of spin in the same sense as the angular velocity of the pendulum about the knife-edge.

MOTION OF A SPINNING BODY

Experiment. — Figure 68 represents a gyroscope attached to a system capable of angular oscillation about a vertical axis. The ring carrying the gyro is incapable of rotation about a vertical axis independently of the outer frame. Set the gyro spinning and set the whole system into angular oscillation about the vertical axis. Observe that the spin-axle sets itself parallel to the axis of oscillation and with the direction of spin in the same sense as the angular velocity of the system.

If a gyroscope be rotated about an axis about which the turning of the gyro-axle is prevented, the gyro-axle tends to set itself parallel to the axis of rotation with the spin of the gyro-wheel in the same sense as the rotation of the gyroscope.

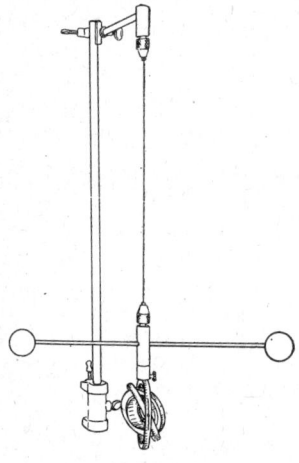

Fig. 68

The same action will occur if one degree of rotational freedom is partially suppressed. Consider a gyroscope with the two gimbal axes horizontal and the spin-axis inclined to the vertical. On rotating the gyroscope about a vertical axis, the spin-axis continues in a fixed direction in space. In order that it may maintain its direction in space, it must rotate with respect to the frame of the gyroscope, about a vertical axis. If rotation about either gimbal axis be opposed by appreciable friction, motion of the spin-axle with respect to the gyroscope frame, about a vertical axis, is opposed by a torque about the vertical axis. When this occurs, the spin-axle revolves about the vertical axis in a converging spiral till it becomes vertical with the direction of spin in the direction of the torque. When there is appreciable friction at one or both of the gimbal axes and the gyroscope is rotated about a vertical axis, the equilibrium position of the spin-axle is vertical whether the center of mass of the gyro is above the intersection of the gimbal axes, is at the intersection, or is below it. If the apparatus is on an airplane making a turn about a vertical axis, the spin-axle will set itself in the direction of the apparent vertical. The action of this apparatus is the basis of devices* for producing a vertical line of reference on moving bodies.

* U. S. Patent. J. and J. G. Gray, No. 1308783, 1919.

MOTION OF A SPINNING BODY

50. The Pioneer Turn Indicator. — Several types of turn indicators are based on the principle that if a gyroscope be rotated about an axis about which turning of the gyro-axle is prevented, the axle will tend to set itself parallel to the axis of rotation with the spin of the gyro-wheel in the same sense as the rotation of the gyroscope.

The Pioneer turn indicator,* widely used by aviators, consists of a gyro-wheel spinning about a horizontal axis transverse to the body of the airplane. The gyro-frame is capable of rotation about the fore-and-aft axis of the airplane AB, Fig. 69. It cannot rotate, relative to the airplane, about a vertical axis.

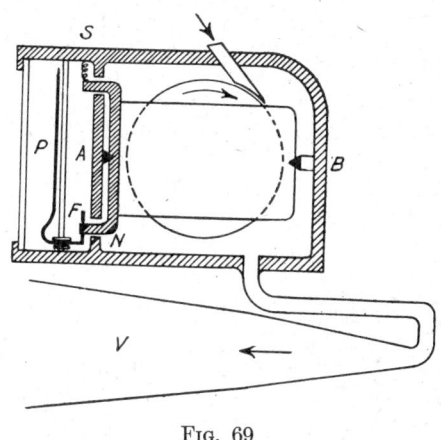

Fig. 69

If the airplane turns about a vertical axis, the gyro-axle tends to set itself vertical with the direction of spin in the direction the airplane is turning about the vertical axis. This turning is opposed by a spring S. The resulting turning of the gyro-frame about the fore-and-aft axis of the airplane is transmitted to a pointer P by means of a pin N and fork F. A nozzle directed toward the row of blades is open to the outside air. The case is exhausted of air by a tube connected to a Venturi nozzle V attached to the outside of the airplane body.

Fig. 70

51. The Clinging of a Spinning Body to a Guide in Contact with it. *Experiment.* — The axle of the Maxwell Top shown in Fig. 70, is adjusted till the point of support coincides with the center of mass of the top. This is done so that gravitational forces need not be considered. Keeping the axle vertical by means of a stick with a round hole in one end, set the top spinning in the usual manner by means of a string. Withdraw the stick, leaving the axle vertical. The top spins steadily

* U. S. Patent. Colvin, No. 1660152, 1928.

about the vertical axis. Pull the upper end of the axle over into contact with the guide G, and then release the axle. Now the end of the axle rolls or slides around against one side of the guide, clinging tightly to it, dashes around the end of the guide, still clinging to it, and rolls or slides along the other side of the guide.

Suppose that the top has an angular momentum about the spin-axis of constant magnitude. When the spinning axle is brought into contact with the guide there is developed on the axle a force of friction represented by the arrow f, Fig. 71. This force, in com-

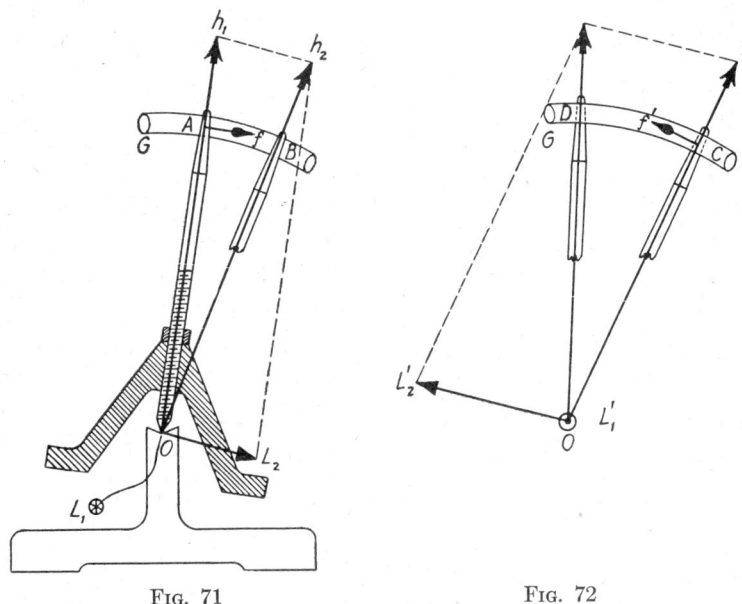

FIG. 71   FIG. 72

bination with an equal, parallel, and oppositely directed force acting on the lower end of the axle, constitutes a torque L_1. Under the action of this torque, the axle precesses against the guide and pushes against it with considerable force. The reaction of this force against the rotating axle, in combination with an equal, parallel, and oppositely directed force acting on the lower end of the axle, constitutes a second torque L_2. Precession developed by this torque causes the upper end of the axle to slide along the guide from the position A toward the position B.

When the axle is on the opposite side of the guide, the friction against the guide develops a torque acting on the axle represented by L_1', Fig. 72. The axle presses against the guide. The reaction of this force develops a torque acting on the axle represented by L_2'. The spin-axle precesses from the position OC toward the position OD.

In this Article we have considered a spinning body with the fixed point coinciding with the center of mass of the body. Pressure is exerted also against the guide when the fixed point of the spinning body is either below or above the center of mass. An important application of this last case is made in the Griffin pulverizing mill, Art. 54.

52. Components of the Torque Acting upon a Spinning Body Having one Fixed Point, Relative to the Three Coördinate Axes of a Rotating Frame of Reference. — In Fig. 73, the line h repre-

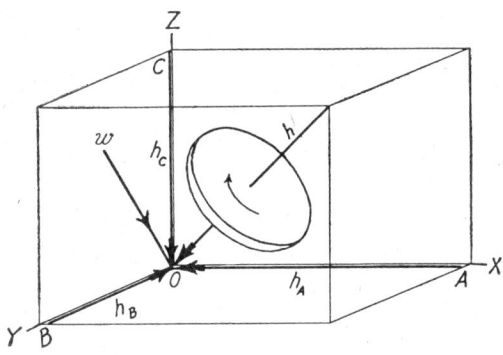

Fig. 73

sents the angular momentum of a spinning body. The axis of this angular momentum passes through the origin of a system of coordinate axes $OABC$. The components of h about the axes of this reference frame are h_A, h_B and h_C, respectively. Let this reference frame $OABC$ be capable of rotation with angular velocity w about an axis through the origin O of a set of rectangular axes $OXYZ$, fixed in space. Denote the projections of w on the axes of the rotatable frame by w_A, w_B and w_C, respectively. It is required to find expressions for the torque, that is the rate of change of the angular momentum of the given body, at any instant, with respect to the moving axes OA, OB and OC.

The torque acting on the spinning body about the axis OA, at any instant, is made up of the sum of four parts — the rate of change of the angular momentum about the axis OA of the angular momentum h if the movable frame were at rest relative to the fixed frame of reference, and the three torques about the axis OA

contributed by the component angular velocities w_A, w_B and w_C of the rotatable frame of reference. Since the components of the angular momentum of the spinning body about the three rotatable axes do not depend upon the position of the fixed axes, the rate of change of these quantities does not depend upon the position of the rotatable frame relative to the fixed axes of coördinates. The determination of these rates of change will be simplest when the two frames coincide.

We shall now find each of the four parts which together make up the component torque about the axis OA acting upon the spinning body. First, note that if the rotatable frame were at rest, the rate of change about the axis OA of the angular momentum of the spinning body would be $\dfrac{dh_A}{dt}$. Second, note that while the component velocity w_A about the axis AO rotates the angular momenta h_B and h_C about AO, the projections of h_B and h_C are always zero. Consequently, a constant angular velocity w_A contributes zero rate of change of angular momentum about AO.

Third, consider the contribution of rate of change of angular momentum about the axis AO, due to the component angular velocity w_C of the movable reference frame. Imagine the component of h in the plane of AOB, Fig. 73, and represented by the line h', Fig. 74, to be rotating with angular velocity w_C about an axis through O, perpendicular to the plane XOY. When h' makes an angle ϕ with AO, the component of h' about AO is $h' \cos \phi$. The rate of change of this component of the angular momentum relative to the axis AO is

$$-h' \sin \phi \frac{d\phi}{dt} = -h' \sin \phi \cdot w_C$$

When $\phi = 0$, this rate of change equals zero, that is, the rotation of h_A contributes zero effect. When $\phi = 90°$, $\sin \phi = 1$, $h' = h_B$; that is, the rate of change of the angular momentum about the axis AO contributed by the rotation of h_B equals $-h_B w_C$. This is the entire rate of change relative to OA due to w_C.

Fourth, consider the rate of change of angular momentum about the axis AO, due to the component angular velocity w_B of the movable reference frame. Imagine the component of h in the plane of AOC, Fig. 73, and represented by the line h'', Fig. 75, to be rotating with angular velocity w_B about an axis through O, perpendicular to the plane XOZ. When h'' makes an angle ϕ

with AO, the component of h'' about AO is $h'' \cos \phi$. The rate of change of this component of the angular momentum relative to the axis AO is

$$-h'' \sin \phi \frac{d\phi}{dt} = h'' \sin \phi \cdot w_B$$

When $\phi = 0$, the rate of change equals zero, that is, the rotation of h_A contributes zero effect. When $\phi = 90°$, $\sin \phi = 1$, $h'' = h_C$; that is, the rate of change of the angular momentum about the

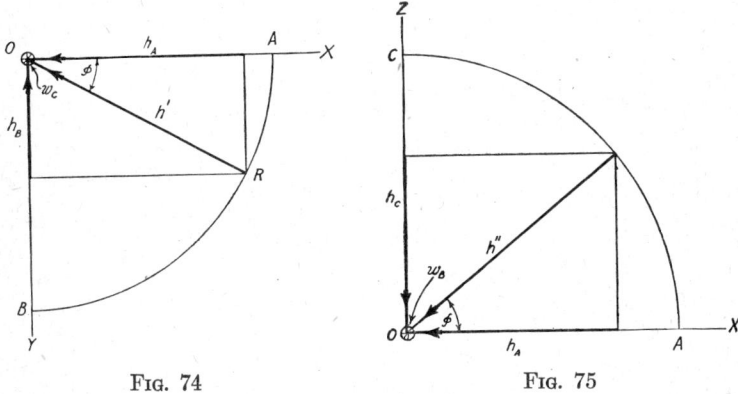

FIG. 74 FIG. 75

axis AO contributed by the rotation of h_C equals $h_C w_B$. This is the entire rate of change relative to OA due to w_B.

Collecting the values of the four torques composing the component acting upon the given body about the axis OA, and representing this component by the symbol L_A, we have,

$$L_A = \frac{dh_A}{dt} - 0 + h_C w_B - h_B w_C \qquad (70)$$

Proceeding in the same manner, we find the component torques acting upon the given body about the axes OB and OC to be, respectively,

$$L_B = \frac{dh_B}{dt} - 0 + h_A w_C - h_C w_A \qquad (71)$$

$$L_C = \frac{dh_C}{dt} - 0 + h_B w_A - h_A w_B \qquad (72)$$

The signs of the last two terms of (70), (71) and (72) may be checked by the following considerations.

With the angular velocity of the movable reference frame $ABCO$, relative to the fixed frame of coördinate axes $XYZO$, as indicated in the diagrams, the rotation of the rigid frame $ABCO$ produces changes in the magnitudes of the component angular momenta about the axes OA, OB, and OC. When the component angular momentum h_B, Fig. 74, is turned with angular velocity w_C, there is a change of angular momentum about the axis OA in the direction opposite h_A. Thus, the rate of change of momentum $h_B w_C$, that is, the torque about the axis OA contributed by the angular velocity w_C is negative. Again, in Fig. 75, the turning of h_C with angular velocity w_B produces angular momentum about the axis OA in the direction of h_A, so that the torque $h_C w_B$ is positive.

Proceeding in the same manner we find that the torque $h_A w_C$ about the axis OB is positive whereas the torque $h_C w_A$ about the same axis is negative. Similarly, the torque $h_B w_A$ about the axis OC can be shown to be positive and the torque $h_A w_B$ to be negative.

When the spin-velocity is constant and about the A-axis, and the precessional velocity is about the B-axis, and the resultant external torque is about the C-axis of a fixed system of coördinate axes $OABC$, then equations (70), (71), and (72) reduce to the expression for the Second Law of Gyrodynamics (55). This is evident if we substitute in (70), (71), and (72),

$$h_A = h_S, \text{ (constant)}, \quad h_C = 0, \quad \text{and} \quad h_B = K_B w_B, \text{ (constant)},$$
$$w_B = w_p \quad \text{and} \quad w_A = w_C = 0$$

53. Components of the Torque Required to Maintain Constant Precession of a Body when the Precession-Axis is Inclined to the Spin-Axis. — Consider a symmetrical body spinning about an axis CO fixed with respect to a rigid frame of reference $OABC$, Fig. 76. The spin-axle CO with the attached rigid reference frame rotates with angular velocity w_z about an axis ZO of a fixed system of rectangular coördinates $OXYZ$. The spin-axle is at a constant inclination θ to the axis ZO. At the instant considered, let AO be in the plane ZOC. Then BO is perpendicular to ZO. The required values of the components with respect to the axes AO, BO, and CO, respectively, of the torque required to maintain a constant velocity of precession of a symmetrical body spinning at constant angular speed w_s about an axis inclined at a constant angle θ to the axis of precession, are obtained from (70), (71), and (72) by substitution of the proper values now to be determined.

Since CO of the moving reference frame rotates with angular velocity w_z about ZO of the fixed system of coördinates, the components of w_z about AO, BO, and CO are, respectively:

$$w_A = -w_z \sin \theta, \quad w_B = 0, \quad \text{and} \quad w_C = w_z \cos \theta \quad (73)$$

Now we shall determine the components of the angular momentum of the spinning body relative to the axes AO, BO, and CO. The component angular velocity of the spinning body about any line equals the sum of the components of w_s and w_z about that line. The component angular velocities of the spinning body relative to the rectangular axes AO, BO, and CO, are $(0 - w_z \sin \theta)$, $(0 + 0)$, and $(w_s + w_z \cos \theta)$, respectively. Representing the moments of inertia of the spinning body relative to the axes AO, BO, and CO by K_A, K_B, and K_C, respectively, the values of the angular momenta of the spinning body about these axes are seen to be

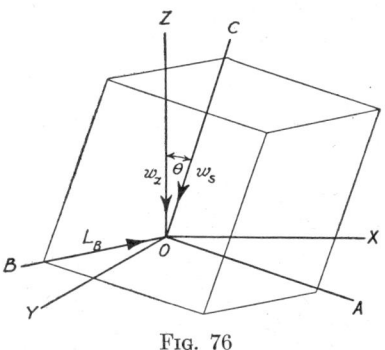

Fig. 76

$$h_A = -K_A w_z \sin \theta, \quad h_B = 0, \quad \text{and} \quad h_C = K_C(w_s + w_z \cos \theta) \quad (74)$$

Assuming that the velocities of spin and precession are constant and the position of the spin-axle relative to the axes AO, BO, and CO, is constant, it follows that

$$\frac{dh_A}{dt} = 0, \quad \frac{dh_B}{dt} = 0, \quad \text{and} \quad \frac{dh_C}{dt} = 0 \quad (75)$$

Substituting in (70), (71), and (72) the values given in (73), (74), and (75), we find that the torques due to the external forces acting upon the spinning body about the axes AO, BO, and CO are, respectively,

$$L_A = 0$$
$$L_B = -K_A w_z^2 \sin \theta \cos \theta + K_C(w_s + w_z \cos \theta) w_z \sin \theta \quad (76)$$
$$L_C = 0$$

Consequently, for steady spin and steady precession about an axis inclined to the axis of spin, there must be zero torque about any line in the plane of the spin and precession axes, but there must be a torque about a line normal to the plane of these axes given by (76).

When the spin-axle is perpendicular to the precession axis, $\theta = 90°$, $\sin \theta = 1$, $\cos \theta = 0$, and (76) becomes the Second Law of Gyrodynamics,

$$L_B = K_C w_s w_z = h_s w_z \qquad (77)$$

When $\theta = 0$, $\sin \theta = 0$, and the torque $L_B = 0$.

The results of Arts. 52 and 53 apply to a top spinning about a fixed point, the Griffin pulverizing mill, and many other cases.

54. The Griffin Pulverizing Mill. — One form of mill for pulverizing cement clinker, ores and other hard materials consists of one or more massive rollers hanging vertically from the ends of a rotating arm. Owing to rotation, each roll presses against the inside of the enclosing pulverizing ring with a centrifugal force proportional to the square of the velocity of the roll relative to the pulverizing ring.

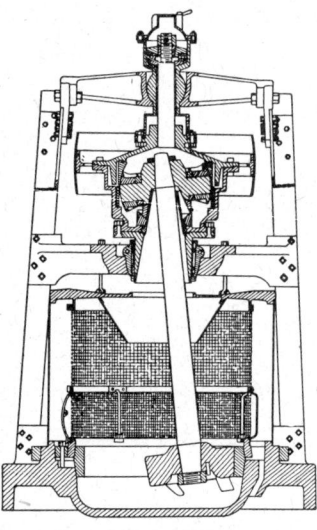

Fig. 77

In the Griffin mill, a roll is hung by a universal joint from the end of a vertical shaft, Fig. 77. If the shaft with the rigidly attached roll be set rotating while its axis is vertical, the axis will remain vertical. If, however, the roll be moved over to the inside of the ring, it will roll around the inner face and push against the pulverizing ring with a centrifugal force and also a force due to gyroscopic torque. Because of the resultant of these two actions the force against the pulverizing ring is greater than it is in the case of the centrifugal roll mill. Paddles on the lower side of the roll toss the material between the roller and the pulverizing ring.

Problem. The roll of a Griffin pulverizing mill weighs 880 lb. and is 8 in. thick. The diameter of the upper face is somewhat greater than that of the lower face and the mean diameter is 23 in. The roll is fastened rigidly on the end of a shaft having a diameter of 5.75 in. and mass of 600 lb. The length of the shaft from the point of suspension to the upper face of the roll is 6 ft. The roll moves around the inside wall of a pulverizing ring having a diameter

ERRATUM

REPLACING THE SOLUTION OF THE PROBLEM, PP. 93-94

of 40 in. Find the force with which the roll presses against the pulverizing ring when the *pulley* is making 165 r.p.m.

Solution. The torque acting upon the roll and shaft due to the rotation about the axis of the shaft, AC, Fig. 78, and the rotation of the axis of the shaft about a fixed vertical axis, OZ, is given by (76). Let

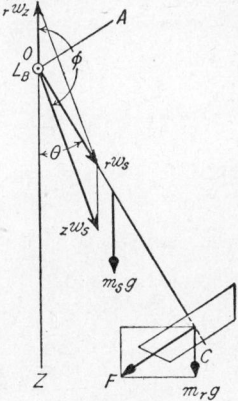

m_r = mass of the roll = $\dfrac{800 \text{ lb.}}{32.1}$ = 27.4 slugs

m_s = mass of the shaft = $\dfrac{600 \text{ lb.}}{32.1}$ = 18.7 slugs

l_r = length of roll = 0.66 ft.
l_s = length of shaft = 6 ft.
d_r = diameter of roll = 1.9 ft.
d_s = diameter of shaft = 0.48 ft.
K_A = moment of inertia of the shaft and roll about an axis, OA, coinciding with the diameter of one end, (24) and (27),

$K_A = m_s \left(\dfrac{d_s^2}{16} + \dfrac{l_s^2}{3} \right) + m_r \left(\dfrac{d_r^2}{16} + \dfrac{l_r^2}{12} + \overline{6.33}^2 \right)$
    = 1331 slug-ft.²

Fig. 78

K_C = moment of inertia of the shaft and roll about the axis of the shaft, OC, (22),
    = $\tfrac{1}{8} m_s d_s^2 + \tfrac{1}{8} m_r d_r^2 = \tfrac{1}{8} (m_s d_s^2 + m_r d_r^2)$
    = $\tfrac{1}{8} (18.7 \times 0.23 + 27.4 \times 3.6) = 12.9$ slug-ft.²

θ = deflection of the roll shaft from the vertical = $\sin^{-1}\left(\dfrac{1.66 - 0.95}{6}\right)$
    = 6° 40'.

$_rw_z$ = angular velocity, about a vertical axis, of the azimuth plane through the roll shaft and the point of suspension, relative to the pulverizing ring. In (76) this is represented by w_z.

$_zw_s$ = angular velocity of the roll shaft relative to the rotating azimuth plane. In (76) this is represented by w_s.

$_rw_p$ = angular velocity of the pulley about its axis, relative to the pulverizing ring = 165 r.p.m.

The value of $_rw_z$ now will be determined. If the azimuthal plane through the pulley shaft and roll shaft were not rotating, then while the roll moves without slipping once around the inside of the pulverizing ring, it would make R/r turns relative to the azimuthal plane. But while the roll makes one circuit of the pulverizing ring, the azimuthal plane makes one revolution in the opposite direction. Thus, the number of revolutions made by a roll of radius r while rolling once, without slipping, around the inside of a ring of radius R is $\left(\dfrac{R}{r} - 1\right)$, relative to the rotating azimuthal plane. Hence, during one revolution of the driving pulley, the roll makes $\dfrac{r}{R - r}$ revolutions relative to the azimuthal plane. The product of this fraction and the angular velocity of the driving pulley, relative to the pulverizing ring, equals the angular velocity of the azimuthal plane relative to the ring,

$$_rw_z \left[= -\frac{r}{R - r} \,_rw_p \right] = -\frac{23}{40 - 23}\, 165 = -223 \text{ r.p.m.}$$
$$= -23.3 \text{ radians per second.}$$

The negative sign is used to indicate that the motion of the azimuthal plane is in the direction opposite that of the driving pulley.

The value of $_zw_s$ now will be determined. While the center of the roll is going once around the pulverizing ring, the roll makes $\dfrac{R}{r}\,_rw_z$ revolutions relative to the azimuthal plane. Hence, the angular velocity of the roll shaft relative to the azimuthal plane

$$_zw_s \left[= \frac{R}{r}\,_rw_z = \frac{R}{R - r}\,_rw_p \right] = \frac{40}{40 - 23}\, 165 = 388 \text{ r.p.m.}$$
$$= 40.6 \text{ radians per second.}$$

Substituting the data and values now found in (76), the magnitude of the torque acting upon the roll and shaft due to the rotations,

$$L_B = -1331\,(-23.3)^2\, 0.12 \times 0.99 + 12.9\,(40.6 - 23.3 \times 0.99)$$
$$(-23.3 \times 0.12) = -86426 \text{ lb.-ft.}$$

The force due to this torque acting on the roll perpendicular to the roll shaft

$$F = \frac{-86426 \text{ lb.-ft.}}{6.33 \text{ ft.}} = -13653 \text{ lb. wt.}$$

The pulverizing ring is acted upon by the horizontal component of the reaction of this force

$$F_r = 13653 \cos 6° 40' = 13517 \text{ lb. wt.}$$

As the roll shaft is deflected from the vertical through an angle θ, there is a horizontal force pulling the roll away from the ring of the value, Fig. 79,

$$f = \frac{(880 \times 6.33 + 600 \times 3.33) \sin \theta}{6.33 \cos \theta} = 142 \text{ lb. wt.}$$

Fig. 79

Consequently, the total force with which the roll pushes against the pulverizing ring is

$$F_r - f = 13517 - 142 = 13375 \text{ lb. wt.}$$

MOTION OF A SPINNING BODY

of 40 in. Find the force with which the roll presses against the pulverizing ring when the roll shaft is making 165 r.p.m.

Solution. The torque acting upon the roll and shaft due to the rotation about the axis of the shaft, AC, Fig. 78, and the rotation of the axis of the shaft about a fixed vertical axis, OZ, is given by (76). Let

m_r = mass of the roll = $\dfrac{800 \text{ lb.}}{32.1}$ = 27.4 slugs

m_s = mass of the shaft = $\dfrac{600 \text{ lb.}}{32.1}$ = 18.7 slugs

l_r = length of roll = 0.66 ft.
l_s = length of shaft = 6 ft.
d_r = diameter of roll = 1.9 ft.
d_s = diameter of shaft = 0.48 ft.
K_A = moment of inertia of the shaft and roll about an axis, OA, coinciding with the diameter of one end, (25) and (27),

$= m_s\left(\dfrac{d_s^2}{16} + \dfrac{l_s^2}{3}\right) + m_r\left(\dfrac{d_r^2}{16} + \dfrac{l_r^2}{3} + 36\right)$

$= 1222$ slug-ft.^{2}

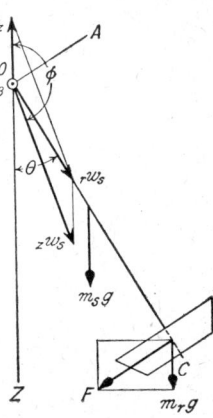

Fig. 78

K_C = moment of inertia of the shaft and roll about the axis of the shaft, OC, (22),

$= \tfrac{1}{8} m_s d_s^2 + \tfrac{1}{8} m_r d_r^2 = \tfrac{1}{8}(m_s d_s^2 + m_r d_r^2)$

$= \tfrac{1}{8}(18.7 \times 0.23 + 27.4 \times 3.6) = 12.9$ slug-ft.^{2}

θ = deflection of the roll shaft from the vertical = $\sin^{-1}\left(\dfrac{1.66 - 0.95}{6}\right)$

$= 6° 40'$.

$_rw_z$ = angular velocity, about a vertical axis, of the azimuth plane through the roll shaft and the point of suspension, relative to the pulverizing ring. In (76) this is represented by w_z.

$_zw_s$ = angular velocity of the roll shaft relative to the rotating azimuth plane. In (76) this is represented by w_s.

$_rw_s$ = angular velocity of the roll shaft about its axis, relative to the pulverizing ring

$= \dfrac{165}{60} 2\pi = 17.2$ radians per second.

The values of $_rw_z$ and $_zw_s$ now will be determined. The number of revolutions made about the axis of the roll shaft by a roll of radius r while rolling once, without slipping, against the inside of a ring of radius R is

$$n = \dfrac{R}{r} - 1$$

Therefore,

$$_rw_z\left[= -\dfrac{_rw_s}{n}\right] = -\dfrac{r \, _rw_s}{R - r} = -23 \text{ radians per sec.} \tag{78}$$

The negative sign is used to indicate that $_rw_z$ is in the sense opposite that of $_rw_s$.

The resultant of the angular velocities $_rw_z$ and $_zw_s$ is $_rw_s$. (See Arts. 4 and 6.) From Fig. 78 and the parallelogram law, (Art. 3),

$$_rw_s^2 = {_rw_z^2} + 2\,_rw_z\,_zw_s \cos\phi + {_zw_s^2}$$

Substituting in this equation the data of the problem as well as the value of $_rw_z$ from (78), and solving the resulting equation for $_zw_s$,

$$_zw_s = 15.5 \text{ radians per sec.} \tag{79}$$

Substituting the values now found in (76), the value of the torque acting upon the roll and shaft due to the rotations

$$L_B = -1222(-23)^2 0.12 \times 0.99 + 12.9(15.5 - 23 \times 0.99)(-23 \times 0.12)$$
$$= -76450 \text{ lb.-ft.}$$

The force due to this torque acting on the roll perpendicular to the roll shaft

$$F = \frac{-76450 \text{ lb.-ft.}}{6.33 \text{ ft.}} = -12077 \text{ lb. wt.}$$

The pulverizing ring is acted upon by the horizontal component of the reaction of this force

$$F_r = 12077 \cos 6° 40' = 11956 \text{ lb. wt.}$$

As the roll shaft is deflected from the vertical through an angle θ, there is a horizontal force pulling the roll away from the ring of the value, Fig. 79,

$$f = \frac{(880 \times 6.33 + 600 \times 3.33) \sin\theta}{6.33 \cos\theta} = 142 \text{ lb. wt.}$$

Consequently, the total force with which the roll pushes against the pulverizing ring is

$$F_r - f = 11956 - 142 = 11814 \text{ lb. wt.}$$

Fig. 79

55. The Automobile Torpedo. — A torpedo that is operated by its own power and is steered automatically by a self-contained apparatus is called an automobile torpedo. We have torpedoes 22 feet long and 21 inches in diameter that have a speed of 40 knots when submerged and have an effective range of 8000 yards. Such a machine consists of some 3000 parts, costs in the neighborhood of $16,000 and is capable of destroying the largest battleship.

An automatically operating gyroscopic steering device can be set so that after the torpedo leaves the vessel the torpedo will maintain an assigned course which may be either straight or bent. The torpedo can be maintained at an assigned distance below the surface of the sea and caused to sink if it does not strike a target.

All modern automobile torpedoes are operated by compressed gas. The large ones are equipped with two engines of some 700

MOTION OF A SPINNING BODY 95

horse-power, each operating a propeller. The two propellers revolve with equal speed in opposite directions on the after ends of two concentric tubular shafts. The exhaust gases escape through the inner tubular shaft. Some European countries use reciprocating engines, usually of four cylinders arranged radially in a plane perpendicular to the shaft.* The U. S. Navy uses rotary engines.

A reservoir some 12 feet long contains air at a pressure of about 2800 pounds per square inch. After being reduced in pressure to

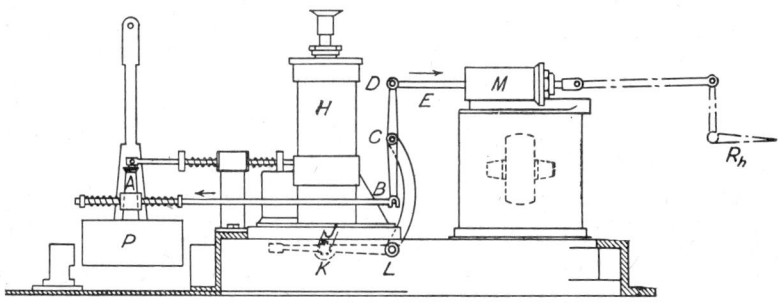

Fig. 80

about 350 pounds per square inch, the air is led into a combustion chamber or " pot " where the temperature is raised by a flame of alcohol or gasoline sprayed into the pot. The fuel spray is ignited by a cordite fuse which in turn is ignited by the explosion of a percussion cap on the launching of the torpedo. The temperature of the pot and contents is prevented from rising too high by spraying into this pot a stream of pure water in addition to the air and fuel. The water flashes into superheated steam thereby adding considerable energy to the working substance.

Gas from the pot operates not only the main motors but also the two steering motors and the turbine that starts the gyro and the one that maintains the spinning.

56. The Pendulum-Controlled Depth Steering Gear. — A horizontal rudder R_h, Fig. 80, operated by piston in a cylinder M maintains the torpedo horizontal at a predetermined depth in the water. The displacement of the piston is controlled by a valve which in turn is controlled by the joint action of a pendulum P and two connected hydrostats, one on either side of the propeller shaft.

* Dumas, " La Torpille Automobile," Le Génie Civil, p. 401 (1915).

Each hydrostat H has a diaphragm J at the lower end of a vertical cylinder. The under side is exposed to the pressure of the sea. Any desired pressure is applied to the upper side by adjusting the tension of a spiral spring that fills the vertical cylinder. If the torpedo sinks below the depth corresponding to the pressure for which the spring is set, the sea pressure pushes the diaphragm upward and the attached bell crank KLC rotates in the clockwise direction about a fixed shaft L thereby moving the valve rod DE in the direction indicated in the figure.

If the nose of the torpedo dips, the pendulum pulls the rod AB in the direction indicated by the adjacent arrow, and the valve-rod DE is pushed in the direction indicated in the figure.

57. The Conditions that Must Be Fulfilled by the Horizontal Steering Mechanism. — No device connected with the automobile torpedo has required so much study and experimentation as that employed to maintain a fixed course however the torpedo may be buffeted by waves. All practical devices depend upon the " rigidity of the spin-axis " in space of a gyro of three degrees of rotational freedom so long as the gyroscope is unacted upon by any outside torque.

The gyro must be mounted in gimbal rings. There must be a mechanism to keep the spin-axle pointing in a predetermined direction so long as the torpedo is within the launching tube. There must be a starting motor that will get the speed of the gyro up to a high value by the time the torpedo leaves the launching tube. There must be a device that will disconnect the locking device as soon as the torpedo enters the water and that will at that instant substitute for the starting motor another motor which will maintain the spin-velocity without interfering with the movement of the gyro-axle relative to the torpedo. A torque must be applied to the gyro that will produce a precessional velocity of the gyro-axle equal and opposite to the angular velocity of the gyro-axle relative to the earth (Art. 41).

While the torpedo is within the launching tube, the gyro of the locked gyroscope is speeded up to about 10,000 revolutions per minute by means of a separate turbine motor geared to the gyro-axle. The torpedo is projected by a charge of cordite. About the time the torpedo strikes the water, the gyroscope is unlocked, the starting motor is disconnected, and two streams of high pressure gas are directed tangentially into shallow buckets cut into the edge of the gyro.

If the torpedo axis is deflected from the direction it had when the gyroscope was unlocked, the angle between the torpedo axis and the gyro spin-axis is changed. The displacement of the gyro spin-axle relative to the torpedo axis is employed to operate a valve which controls the passage of compressed gas to one side or the other of the piston of a steering motor connected to the vertical rudders. In the transmission of the considerable force required to operate this valve, no appreciable torque must be applied to the gyroscope. Otherwise, the gyro-axle would be caused to precess and no longer remain in a fixed direction in space.

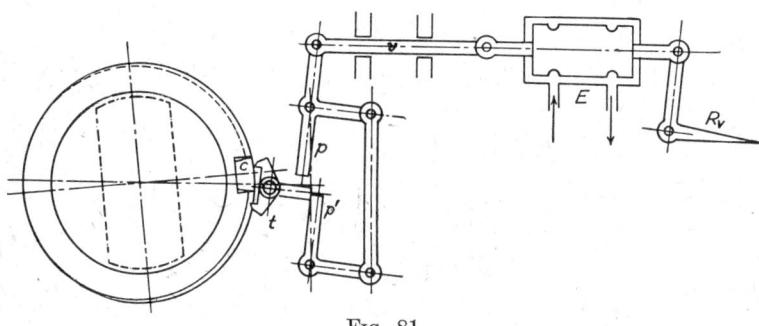

Fig. 81

58. The Bliss-Leavitt Torpedo Steering Gear. — These conditions have been met in the Bliss-Leavitt torpedo steering mechanism based on the patents of F. M. Leavitt and Wm. Dieter.*
The Bliss-Leavitt torpedo is used in the United States and other navies.

The operation of the device can be understood from an inspection of Fig. 81, which is a simplified plan view of the mechanism for controlling the vertical rudders. The spin-axis of the gyro is horizontal, Fig. 82, the axis of rotation of the inner gimbal ring is horizontal and perpendicular to the spin-axis; the axis of rotation of the outer gimbal ring is vertical. Attached rigidly to the upper side of the

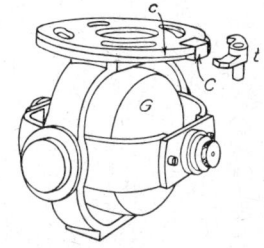

Fig. 82

---

* U. S. Patents. Leavitt, No. 741683, 1903; No. 768291, 1904; No. 795045, 1905; No. 785424, 1905; No. 814969, 1906; No. 901355, 1908; No. 925709, 1909; No. 925710, 1909; No. 1080116, 1913; No. 1145025, 1915; No. 1197134, 1916; No. 1291031, 1919.
Dieter, No. 1148154, 1915; No. 1233761, 1917; No. 1318980, 1919; No. 1402745, 1922; No. 1440822, 1923.

outer gimbal ring is a horizontal disk having a cam on the edge. This cam comprises a rectangular part C, Fig. 82, concentric with the edge of the disk, and a spiral part c, like one turn of the square thread of a jackscrew. A light tappet, t, capable of angular motion about a vertical pivot, oscillates rapidly toward and away from the cam along the radius of the cam-disk parallel to the longitudinal axis of the torpedo. The tappet has two fingers at different levels.

Normally, that is when the torpedo is on a straight course, the axis of the torpedo and the radius of the cam-disk through the

Fig. 83

center of the square part of the cam are parallel to the gyro spin-axle. In this case, when the tappet moves up to the cam-disk, the two fingers strike the square part C of the cam and the tappet will be turned till its axis is parallel to the radius of the cam-disk that goes through the center of the square part of the cam. This position of the tappet is preserved during the reverse stroke. On the reverse stroke, the arm of the tappet will move into the gap between the two pallets p and p', Fig. 81. Just as soon, however, as the torpedo axis departs slightly from parallelism to the gyro spin-axis, one finger of the oscillating tappet strikes the cam and the other strikes the edge of the disk. As the cam is farther from the center of the disk than is the edge of the disk, the tappet will be tilted to the left or right of the torpedo axis. On the reverse stroke, the arm of the tappet will strike one of the pallets p or p', tilt the two connected elbow levers and move the valve rod v either forward or backward. The displacement of the valve rod allows compressed gas from the combustion pot to enter the steering motor E on one side of the piston or the other, thereby turning the vertical rudder either hard to port or hard to starboard. Figure 83 gives a view of the assembled control mechanism.

59. Method of Compensating the Effect of the Rotation of the Earth. — During the time that a torpedo is making a run of ten minutes, the earth rotates 2.5°. If the gyro-axle remains fixed in space, its direction relative to the earth changes by an amount that depends upon the latitude and also upon the direction in which the torpedo is moving. The deflection of the spin-axle in azimuth can be compensated by the application of a torque that will maintain a precessional velocity equal and opposite to the proper component of the angular velocity of the earth. The required torque is produced by the weight of a small nut that can be moved along a screw fastened to the inner gimbal ring and extending in the direction of the gyro-axle. To obtain the correct torque, the gyroscope is mounted on a stand, the gyro set spinning at the correct speed, and the position of the adjusting nut changed till the gyro-axle remains stationary in azimuth. This adjustment also compensates for any lack of balance of the gyroscope that would produce a precession.

The spin-velocity of a torpedo gyro is not constant throughout a long run. The precessional velocity due to any given torque varies inversely with the spin-velocity (57). Consequently, the adjusting nut is placed so that the mean precession of the gyro equals the angular velocity of the gyro-axle due to the rotation of the earth at the given place. If the adjustment is correct for a ten-minute run at a given latitude, then for a run of shorter duration, or at a place nearer the pole, the torpedo will be deflected to the right.

60. Devices for Changing the Course of a Torpedo. — Several methods have been devised for causing an automobile torpedo to make a turn of any predetermined angle and thereafter proceed along a straight course. The Leavitt* method of " angle fire " or " curved fire " is to turn the cam-disk from its zero position on the second frame of the gyroscope through the required angle in the proper direction. The rudder is thereby held hard over till it

* U. S. Patent. Leavitt, No. 925710, 1909.

has steered the torpedo through the predetermined angle, that is till the line from the center of the cam-disk to the center of the square part of the cam is parallel to the axis of the torpedo. From that moment, the torpedo will be steered in a straight course.

The Dieter* method of angle fire is to keep the cam-disk fastened to the gyroscope so that the line from the center of the disk to the center of the square part of the cam remains parallel to the gyro spin-axle, and turn the entire gyroscope through the required angle about an axis perpendicular to the torpedo axes. The operation is similar to that of the preceding method.

Other methods for producing angle fire have been devised by Kaselowsky, Waldron, Patterson, Blount† and others.

A torpedo proceeding along a straight course has but one chance to hit a given target. If, however, after reaching the neighborhood of the target, the torpedo be caused to go around and around in circular paths, then there are more chances of a hit. Devices to cause a torpedo to proceed on a straight course for a predetermined distance and then execute circular paths have been developed by several inventors.‡

61. Airplane Cartography. — During the Great War methods were developed by which military maps were made from a series of photographs taken by a camera mounted on an airplane. As the airplane moved back and forth in parallel paths, the camera took a series of overlapping views on a long film which afterward could be cut up and fastened together so as to form a mosaic of the district covered. The points which were common to overlapping parts of successive pictures served to bring the separate pictures into register and also to show any difference in scale of successive pictures.

Since the war, the method has been greatly improved and its use extended to civil operations. It is estimated that over rough terrain one camera on an airplane can take in one hour the pictures for a reconnaissance map that would be as accurate and detailed as the data that would be taken in one month by a party of one hundred surveyors.

In order that all the pictures in a strip may be on the same

---

* U. S. Patent. Dieter, No. 1153678, 1915.

† U. S. Patents. Kaselowsky, No. 661535, 1900; Waldron, No. 983467, 1911; Patterson, No. 1332302, 1920; Blount, No. 1527777, 1925.

‡ U. S. Patents. Dieter, Nos. 1303038 and 1303044, 1919; Meitner and May, No. 1401628, 1921; Trenor, No. 1517873, 1924; Bevans, No. 1527775, 1925.

scale, the distance between the camera and the ground would need to be constant. It is, however, impossible and unnecessary to maintain this distance constant. Any difference in the scale of adjacent pictures is shown in the overlapping portions. The pictures that are out of scale are enlarged or diminished to the required degree by photography and the new pictures used in the mosaic.

In order that the photographs may be without distortion, the external axis of the camera must be maintained in the direction of the radius of the earth. If the camera be suspended pendulum-wise, the axis will be deflected from the vertical when the velocity of the airplane changes in either magnitude or direction. It is practically impossible to keep the velocity constant under usual atmospheric conditions.

62. Direct Control of the Direction of the Axis of a Camera. — The first aviator photographers had no means to correct the deflection of a camera produced by acceleration of the airplane, except by manual adjustment. The camera was supported pendulously and moved back as soon as a deflection was observed. Some aviator photographers still use the same method. This method, however, brings the camera back to only approximately the correct position after a noticeable deflection has occurred. The pictures taken while the original deflection is occurring and while it is being corrected are distorted.

Probably the most obvious plan to prevent the deflection of the camera is to apply the First Law of Gyrodynamics, by attaching to the camera one or two gyroscopes.* This plan is faulty in that when a gyroscope exerts a torque in erecting a camera, the gyro itself is acted upon by a torque, and this torque causes the spin-axle and the attached camera to turn about an axis perpendicular to the torque-axis. Thus, the camera is given an undesired displacement. Again, even though the spin-axle were directed toward the center of the earth at one instant and the direction of the spin-axle remained fixed in space, the spin-axle will not point to the center of the earth at later instants. Consequently, the pictures taken at later instants will be distorted.

63. Indirect Control of the Direction of the Axis of a Camera. — A camera or other device can be tilted by a torque produced by a motor that is started in either direction, stopped or reversed, by a

* U. S. Patents. Fairchild, No. 1546372, 1925; Lucian, No. 1634950, 1927; Titterington, No. 1645079, 1927.

gyroscope. Fairchild and Morton* have devised an apparatus for this purpose that includes a gyroscope the axis of which continues to point toward the center of the earth. The gyroscope controls two motors without itself being acted upon by sufficient torque to disturb the direction of the spin-axle.

The camera C is fastened rigidly to one end of a frame AA', Fig. 84, capable of rotation about two horizontal axes, perpendicular to one another, through the point O. The casing G of a gyro with vertical spin-axle is supported non-pendulously in gimbal rings on the other end of the frame. The gyro-casing carries a horizontal ring R capable of being turned about the spin-axle. Fast-

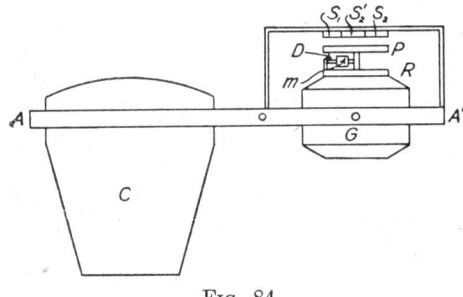

Fig. 84

ened to this ring is a threaded rod D in the direction of a radius of the ring. A mass m can be moved along this rod.

Suppose that the instrument is in the northern hemisphere, the spin-axle is vertical and the gyro is spinning in the clockwise direction as viewed from above. After a time, although the spin-axle will be pointing still in the same direction in space as before, it will be pointing to the east of the center of the earth because the earth has rotated meantime from the west toward the east. The direction of the spin-axle can be maintained vertical by precessing the spin-axle about a horizontal axis in the meridian plane of the earth, with an angular velocity equal to that of the earth and in the opposite direction. The required precessional velocity can be produced by the application of the proper torque about an east-west axis. With the direction of spin as specified above, the desired torque is developed when the mass m is at the proper distance to the north of the spin-axle. By this device the spin-axle is caused to maintain a practically vertical position.

* U. S. Patents. Fairchild and Morton, No. 1559688, 1925; No. 1679354, 1928.

One method by which the gyro may be used to keep the camera axis pointing to the center of the earth involves the use of two motors controlled by currents induced by any motion of the camera relative to the gyro. One motor can turn the frame about an axis AA' and the other about a horizontal axis perpendicular to AA'. A flat coil P, with axis vertical, is fastened to the upper face of the gyro-casing. This is traversed by a high-frequency alternating current of a few milliamperes produced by the rotation of an armature forming part of the gyro. Four other flat coils with axes vertical are supported above the coil P by a bracket fastened to the frame. The five coils are represented in perspective in Fig. 84a.

Fig. 84a

The coils S_1 and S_2 constitute part of a secondary circuit that includes a three-electrode vacuum tube, transformer, condensers, direct current motor and battery. The coils S_1' and S_2' constitute part of another secondary circuit. So long as the primary coil P is equally distant from the four secondary coils, zero electromotive force is induced in each secondary circuit and neither motor starts. This is the condition when the camera is vertical.

If the end A of the frame tilts upward, the system of four secondary coils becomes inclined to the horizontal plane of the coil P, Fig. 84a, coil P now is nearer S_2 than to S_1 and it is equally distant from S_1' and S_2'. An electromotive force is being induced in $S_1 S_2$ whereas zero electromotive force is being induced in $S_1' S_2'$. The motor in the $S_1 S_2$ circuit starts and tilts the frame till the plane of the secondary coils is parallel to the plane of the primary coil P, that is, till the camera axis becomes vertical.

In the same manner, if the frame becomes tilted about an axis AA', the distance between the primary coil and the two secondary coils S_1' and S_2' becomes unequal. The motor in the $S_1' S_2'$ circuit starts and tilts the frame back till the camera axis is again vertical. The currents in the primary and secondary coils are so minute that these actions produce an inappreciable force on the gyro and therefore no precession.

64. Control of the Line of Sight of a Camera. — The camera, instead of being held vertical over the ground to be photographed, may be mounted rigidly along the fore-and-aft axis of the airplane and a vertical beam of light from the ground reflected into the lens system. This can be done by either a plane mirror or a totally reflecting prism placed in front of the lens system. In order that the image formed by light from an object vertically below the airplane may remain fixed in position on the sensitive film when the airplane turns about either a longitudinal or transverse axis, the reflector must be turned in the opposite direction to that in which the airplane turns, and to half the extent. This result can be produced by a gyroscope attached to the reflector.*

* U. S. Patents. Sperry, No. 1688559, 1928; Henderson, No. 1709314, 1929.

CHAPTER III

THE GYROSCOPIC PENDULUM OR PENDULOUS GYROSCOPE

§1. General Properties

65. The Gyro-Pendulum. — A gyroscope mounted so that the center of gravity is either below or above the intersection of two horizontal perpendicular axes about which the system can oscillate is called a gyroscopic pendulum, gyro-pendulum, or pendulous gyroscope. Since a gyro-pendulum with the center of mass below the point of support has greater stability than a compound pendulum of equal mass, it is much used for stabilizing cameras, telescopes and other instruments subject to accelerations on ships and airplanes. Some forms of gyro-compasses and ship stabilizers are pendulous gyroscopes. Inverted gyro-pendulums have been used to stabilize vehicles that are statically unstable such as vehicles designed to operate on a single rail. A gyro-pendulum may be arranged to oscillate in one plane like an ordinary pendulum, or may be arranged to oscillate as a conical pendulum.

A gyroscope fastened to an oscillating body so as to apply a periodic torque to the body, may be arranged in such a manner that the successive vibrations of the oscillating body may be either increased or diminished. Such results depend upon the principle proved in Art. 25. *In case a periodic torque acts upon an oscillating body of the same frequency, (a) energy will be imparted to the oscillating body at the maximum rate, and the amplitude of vibration will increase at the maximum rate, when the torque is in phase with the angular velocity of the oscillating body; (b) energy will be abstracted from the oscillating body, and the amplitude of vibration will diminish at the maximum rate, when the torque is in opposite phase to the angular velocity of the oscillating body.*

66. The Period and the Equivalent Length of a Gyroscopic Conical Pendulum. — In Fig. 85, the center of gravity of the rotating system of weight mg is at a distance l from the point of support C. The gravitational torque L is counter-clockwise about a horizontal axis through C perpendicular to the plane of the diagram. The angular momentum about the spin-axle is represented by h_s. The precessional velocity w_p is about a vertical axis. If the inclination of the spin-axle to the vertical be represented by θ, then the external torque L is given by

$$L = mgl \sin \theta$$

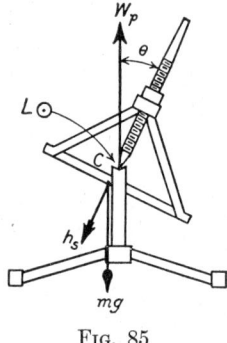

FIG. 85

From the Second Law of Gyrodynamics, Art. (34), this external torque equals the product of the precessional velocity and the component of the angular momentum about an axis perpendicular to the torque-axis and the precession axis. Thus

$$mgl \sin \theta = w_p h_s \sin \theta$$

Hence the precessional velocity

$$w_p = \frac{mgl}{h_s}$$

The period of the precessional motion, that is the period of the gyroscopic conical pendulum

$$T \left[= \frac{2\pi}{w_p} \right] = \frac{2\pi h_s}{mgl} \tag{80}$$

If the radius of gyration of the precessing system, with respect to the spin-axis, be k_s, the above expression may be put into the forms

$$T \left[= \frac{2\pi h_s}{mgl} \right] = \frac{2\pi K_s w_s}{mgl} = \frac{2\pi k_s^2 w_s}{gl} \tag{81}$$

The length of a simple pendulum of the same period, from (51), (80), and (81) is

$$l_e = \frac{T^2 g}{4\pi^2} = \frac{h_s^2}{m^2 l^2 g} = \frac{k_s^4 w_s^2}{gl^2} \tag{82}$$

67. The Inclination of the Precession Axis of a Gyroscopic Conical Pendulum to the Vertical. — A spinning gyro on the earth has two component angular velocities, one w_s, about the spin-axis of the gyro, and another, w_e, about the axis of the earth. The centrifugal couple tending to bring the spin-axis into parallelism with the axis of the earth is about an axis perpendicular to the plane of the spin-axis and the axis of the earth. The axis of the resulting precession is inclined to the vertical at a small angle now to be determined.

In Fig. 86, the point G represents the center of mass of a gyro on the earth at latitude λ; the line AG is the spin-axis of the gyro and D is the point of support; the line VC is the true vertical and HH' is the true horizontal at G; the line NS is parallel to the axis of the earth. The spin-axis is inclined to the axis of the earth at an angle θ.

The gyroscopic torque has the magnitude given in (58). It also equals the product of some force GF parallel to the axis of the earth and a lever arm $DQ[= DG \sin \theta]$.

Or
$$h_s w_e \sin \theta = GF(DG \sin \theta)$$

Whence, the force parallel to the earth's axis has the value

$$GF = \frac{h_s w_e}{l} \qquad (83)$$

where l represents the distance DG.

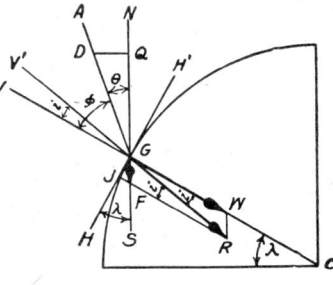

Fig. 86

The gyro is also acted upon by its weight mg, represented in the figure by the line GW. The resultant GR of the two forces GF and GW is the apparent force of gravity acting on the gyro. Its direction is the apparent vertical about which the gyro-axle precesses. The angle i between the true vertical VC and the axis $V'R$ about which the gyro-axle precesses is given by the equation

$$\tan i \left[= \frac{GJ}{JR} \right] = \frac{GF \cos \lambda}{JF + FR}$$

Since i is very small, we may write $\tan i = i$. Since JF is very small, compared to $FR(=GW)$, it may be neglected and the above equation may be written

$$i = \frac{GF \cos \lambda}{mg} \qquad (84)$$

From (83), this becomes

$$i = \frac{h_s w_e \cos \lambda}{mgl}$$

Now the external torque acting on the gyroscope is $mgl \sin \phi$, where ϕ is the angle between the spin-axis and the vertical. The component of the angular momentum about an axis perpendicular

to the vertical is $h_s \sin \phi$. Hence, from the Second Law of Gyrodynamics, (Art. 36),
$$mgl \sin \phi = h_s w_p \sin \phi$$
Substituting the value of mgl from this equation in (84), we obtain for the angle between the precession axis and the vertical

$$i = \frac{w_e \cos \lambda}{w_p} \qquad (85)$$

The angular velocity of the earth

$$w_e \;[= 2\,\pi \text{ radians per day}] = \frac{2\,\pi}{86{,}400} \text{ radians per sec.}$$

and the angular velocity of precession

$$w_p = \frac{2\,\pi}{T} \text{ radians per sec.,}$$

where T represents the period of precession.

Substituting in (85) these values of w_e and w_p, and remembering that
1 radian = 3438 minutes of angle,

$$i = \frac{T \cos \lambda}{86{,}400} \text{ radians} \left[= \frac{T \cos \lambda \,(3438)}{86{,}400} \text{ minutes} \right]$$

$$= \frac{T \cos \lambda}{25.14} \text{ minutes of arc} \qquad (86)$$

Thus, when the period of precession is 25.14 seconds of time, the angle i at the equator is one minute of arc.

68. The Period of the Undamped Vibration, Back and Forth Through the Meridian, of the Gyro-Axle of a Pendulous Gyroscope. — Figure 87 represents the gyro-axle AO of a pendulous gyroscope referred to rectangular coördinate axes OV, ON, and OE that extend vertically upward, northward and eastward. The gyro-axis is in the vertical plane VOH. It is inclined to the meridian plane VON at the angle ϕ and to the horizontal plane EON at the angle θ. The center of mass of the gyro and frame is at G and the point of suspension of the moving system is at D. The distance DG will be represented by the symbol l. The gyro-axis AO and the line DG are perpendicu-

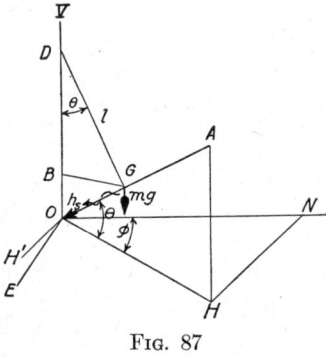

FIG. 87

lar to one another. The weight mg of the moving system has a lever arm BG about an axis through D perpendicular to the plane VOH and produces a torque about this axis of the value $L = mgl \sin \theta$.

Representing the angular velocity of the moving system about a vertical axis OV fixed in space by w', and the angular momentum of the gyro-wheel with respect to a horizontal axis OH by h_s', then when the precession is steady we have

$$L = h_s' w'$$

Since the velocity of precession of the moving system about a vertical axis fixed to the earth is $d\phi/dt$, the velocity of precession about a vertical axis fixed in space equals the sum of $d\phi/dt$ and the component velocity of the earth about the vertical at the given place. If w_e represents the angular velocity of the earth about its axis, then the component angular velocity OD, Fig. 88, about a vertical axis through a point D at latitude λ is $w_e \sin \lambda$. Hence the velocity of precession about a vertical axis fixed in space is

$$w' = \frac{d\phi}{dt} + w_e \sin \lambda$$

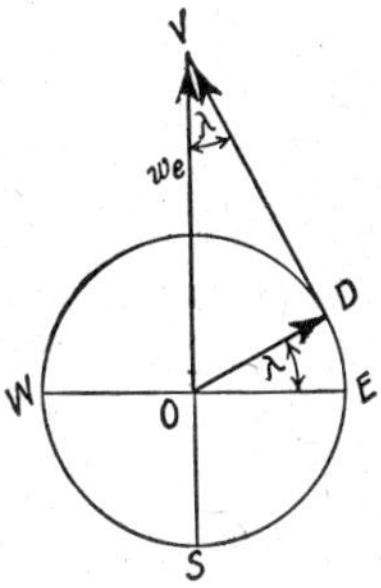

Fig. 88

The component of h_s about OH perpendicular to OV, Fig. 87, is

$$h_s' = h_s \cos \theta$$

Consequently, for an axis fixed in space through D and perpendicular to the vertical plane VOH,

$$[L =] mgl \sin \theta [= h_s' w'] = h_s \cos \theta \left(\frac{d\phi}{dt} + w_e \sin \lambda \right)$$

When θ is so small that the difference between $\cos \theta$ and unity may be neglected and also the difference between $\sin \theta$ and θ radians, we may write

$$mgl\, \theta = h_s \left(\frac{d\phi}{dt} + w_e \sin \lambda \right) \qquad (87)$$

The left-hand side of this equation is the gravitational torque which tends to cause the gyro-axle to become horizontal. The right-hand side is the gyroscopic resistance offered by the gyro-axle opposing the tendency to dip.

Again, the velocity of precession about a horizontal axis OH', or NH, Fig. 87, fixed to the earth and perpendicular to the spin-axis OA, is $d\theta/dt$. The component angular velocity of the earth about a horizontal axis DV, Fig. 88, in the meridian plane, is $w_e \cos \lambda$. This is the component angular velocity of the earth about ON, Fig. 87. The component of this about OH' or the parallel line NH is $w_e \cos \lambda \sin \phi$. Hence, the velocity of precession about a horizontal line OH' fixed in space is

$$\frac{d\theta}{dt} + w_e \cos \lambda \sin \phi$$

The torque about the axis perpendicular to OH' and OA, that is about the axis DG, is zero. Consequently,

$$0 = h_s \left(\frac{d\theta}{dt} + w_e \cos \lambda \sin \phi \right)$$

Differentiating (87) we have

$$mgl \frac{d\theta}{dt} = h_s \frac{d^2\phi}{dt^2}$$

Here are two equations, one obtained by considering precession about a vertical axis and the other by considering precession about a horizontal axis. Each equation involves both θ and ϕ. Now we shall eliminate θ and from the resulting equation study the motion of the spin-axis in the horizontal plane.

On combining the last two equations by eliminating $\frac{d\theta}{dt}$, we find

$$\frac{h_s}{mgl} \frac{d^2\phi}{dt^2} = -w_e \cos \lambda \sin \phi$$

which can be put into the form

$$\frac{d^2\phi}{dt^2} = -\left(\frac{mgl}{h_s} w_e \cos \lambda \right) \sin \phi$$

In order to abbreviate the labor of repeatedly copying the quantity within the parenthesis, we will represent it by the symbol A. When ϕ is small, $\sin \phi = \phi$ radians and the above equation assumes the form

$$\frac{d^2\phi}{dt^2} = -A\phi$$

Multiplying both sides by $2 \frac{d\phi}{dt}$ and integrating, we obtain

$$\left(\frac{d\phi}{dt} \right)^2 = -A\phi^2 + C$$

For convenience in subsequent integration, let $C = AB^2$. Then

$$\left(\frac{d\phi}{dt}\right)^2 = A(B^2 - \phi^2)$$

Extracting the square root,

$$\frac{d\phi}{dt} = \sqrt{A}\sqrt{B^2 - \phi^2} \quad \text{or} \quad \frac{d\phi}{\sqrt{B^2 - \phi^2}} = dt\sqrt{A}.$$

Measuring time from the instant when $\phi = 0$, that is, when the gyro-axle is in the meridian, and integrating the last equation, we obtain

$$\sin^{-1}\frac{\phi}{B} = t\sqrt{A} + C_1$$

When $t = 0$, $\phi = 0$ and hence the first member is zero and $C_1 = 0$. Whence

$$\phi = B \sin t\sqrt{A}$$

which is of the same form as the equation of a simple harmonic motion of rotation, (48),

$$\phi = \Phi \sin\left(t\frac{2\pi}{T}\right)$$

Therefore, the undamped motion of the gyro-axle of a pendulous gyroscope back and forth across the meridian is approximately simple harmonic and of the period $T = \dfrac{2\pi}{\sqrt{A}}$.

Replacing the value of A in this equation

$$T = \frac{2\pi}{\sqrt{\dfrac{mglw_e \cos\lambda}{h_s}}} = 2\pi\sqrt{\frac{h_s}{mglw_e \cos\lambda}} \tag{88}$$

Among other things, this equation shows that the period of the gyro-axle of a pendulous gyroscope back and forth across the meridian is increased by diminishing the distance between the point of support and the center of gravity. This fact is utilized in the design of various gyro-apparatus.

69. The Torque with Which the Second Frame of a Gyroscope Resists Angular Deflection. — Consider a gyro-wheel G of spin-velocity w_s and moment of inertia K_s with respect to the spin-axis, mounted in two coplanar frames A and B, Fig. 89. Let the axis pp' pass through the center of gravity of the wheel and inner frame.

112 THE GYROSCOPIC PENDULUM

If the outer frame be turned through a small angle θ with angular velocity $\frac{d\theta}{dt}$ about an axis at C normal to the plane of the frames while the gyro-wheel is spinning and the two frames are clamped together, the inner frame will be acted upon by a torque about the axis pp' equal to $K_s w_s \frac{d\theta}{dt}$.

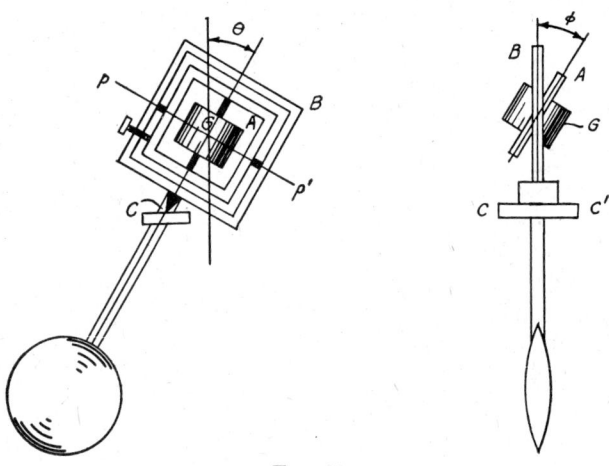

Fig. 89

If the two frames are not clamped together, the inner frame will precess through an angle ϕ in time t, about the axis pp', with an angular acceleration $\frac{d^2\phi}{dt^2}$. If ϕ remains small, we have from (19) and (58)

$$K_p \frac{d^2\phi}{dt^2} = K_s w_s \frac{d\theta}{dt}$$

where K_p is the moment of inertia of the inner frame together with the gyro-wheel, with respect to the axis pp' about which the inner frame is turning.

Let us measure angles and time from the instant when the two frames are in the same plane. At this instant $t = 0$ and $\theta = 0$. Integrating the above expression between the limits 0 and t, we have, when θ is small,

$$K_p \frac{d\phi}{dt} = K_s w_s \theta + 0$$

or

$$\frac{d\phi}{dt} = \frac{K_s w_s \theta}{K_p} \tag{89}$$

If the inner frame were clamped so that the gyro could not precess, a certain torque would be required to give the apparatus a chosen angular acceleration. When the gyro is precessing, the centrifugal couple must be balanced by an additional torque. In order to give the apparatus the chosen angular acceleration it is necessary to apply an outside torque which is the sum of these two components. That is,

$$L_c = K_c \frac{d^2\theta}{dt^2} + K_s w_s \frac{d\phi}{dt}$$

Substituting in this equation the value of $\frac{d\phi}{dt}$, found above,

$$L_c = K_c \frac{d^2\theta}{dt^2} + \frac{K_s{}^2 w_s{}^2 \theta}{K_p} \qquad (90)$$

This is the torque about an axis normal to the precession axis that is required to turn the inner frame together with the gyro-wheel with an angular acceleration.

When the inner frame with the gyro-wheel is turned with constant velocity, the first term in the right-hand member of the above equation is zero. Consequently, when the gyro-axle of a precessing gyro-wheel is turned with constant angular velocity through a small angle θ about an axis perpendicular to the precession axis, the precessing wheel exerts a torque on the restraints of the value

$$L_c' = \frac{K_s{}^2 w_s{}^2 \theta}{K_p} = \frac{h_s{}^2 \theta}{K_p} \qquad (91)$$

where K_p is the moment of inertia of the gyroscope with respect to the precession axis.

70. The Length of the Simple Pendulum That Has the Same Period as an Oscillating Body to Which is Attached a Spinning Gyroscope. — When the gyroscope is not spinning, the pendulous body has a period given by (50). A simple pendulum has a period given by (51). If the simple pendulum has the same period as the pendulous body,

$$l = \frac{K_c}{mH} = \frac{k^2}{H} \qquad (92)$$

where H represents the distance from the center of mass of the pendulous body to the knife-edge, m is the mass of the pendulous body, K_c is the moment of inertia of the pendulous body with respect to the axis of oscillation, and l is the length of the simple pendulum that has the same period as the given pendulous body.

114 THE GYROSCOPIC PENDULUM

When the pendulous body is inclined to the vertical at an angle θ, the total torque $K_c \mathbf{a}_1$ acting upon it is due to gravitational forces and has the value, (19 and 49):

$$K_c \mathbf{a}_1 = -mgH \sin \theta$$

where the negative sign indicates that angular displacements are in the direction opposite the torque. When θ is so small that $\sin \theta$ may be replaced by θ radians, the total torque about the knife-edge

$$K_c \mathbf{a}_1 \, [= mk^2 \mathbf{a}_1] = -mgH\theta \qquad (93)$$

or the magnitude of

$$mH = \frac{K_c \mathbf{a}_1}{g\theta}$$

Substituting in (92) this value of mH, we find the length of the simple pendulum that has the same period as the oscillating body to which is attached an unspinning gyroscope to be

$$l_1 = \frac{g\theta}{\mathbf{a}_1} = \frac{mk^2}{mH} \qquad (94)$$

When the gyro is spinning and the spin-axle is precessing about a horizontal axis perpendicular to the knife-edge, the total torque acting on the pendulum about the knife-edge is made up of two parts, one due to gravity and another due to precession. The torque due to precession equals $h_s w_p$. From (89)

$$w_p \left[= \frac{d\phi}{dt} \right] = \frac{h_s \theta}{K_c}$$

where ϕ represents the angle of precession when the pendulum is deflected θ from the equilibrium position. Hence, the total torque acting on the pendulum about the knife-edge, when the spinning gyro is precessing about a horizontal axis perpendicular to the knife-edge, has the value

$$K_c \mathbf{a}_2 = -mgH\theta - \frac{h_s \theta}{K_c} = -g\theta \left(mH + \frac{h_s^2}{gK_c} \right) \qquad (95)$$

where $\mathbf{a}_2$ represents the angular acceleration of the system and K_c represents the moment of inertia, both with respect to the knife-edge. The negative signs indicate that θ is measured in the direction opposite the angular acceleration.

Whence, the length l_2 of the simple pendulum which has the same period as a statically stable body that is carrying a spinning and precessing gyroscope has the value (94 and 95)

$$l_2 \left[= \frac{g\theta}{a_2} \right] = \frac{mk^2}{mH + \dfrac{h_s^2}{gK_c}} \qquad (96)$$

A comparison of (94) with (96) shows that when the attached gyroscope is spinning, the length of the simple pendulum equivalent to the given gyro-pendulum becomes shorter than when the gyro is not spinning, that is the period of vibration of the gyro-pendulum becomes shorter.

§2. *Gyro-Horizontals and Gyro-Verticals*

71. Determination of the Latitude of a Place. — The latitude of a place is the angular distance of the place from the equator, measured on a meridian. It is expressed in degrees, minutes, and seconds north or south of the equator. In Fig. 90, EQ represents the equator of the earth and NS the polar axis. HO is the horizontal at the point X. The angle XCE is the latitude of X.

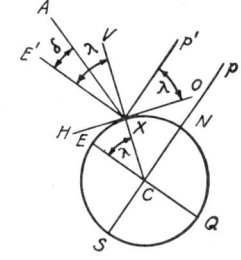

Fig. 90

The latitude can be found from an observation of the altitude above the horizon of the sun, a planet, or a star. The sun crosses the meridian at noon. Suppose that at this time the light from the sun to the point X follows the line AX, and that the declination of the sun, that is the angular distance of the sun above the celestial equator, is δ. Since the altitude of the sun is AXH, Fig. 90, we see that the latitude of X is given by

$$\lambda = 90° - AXH + \delta$$

Altitudes of heavenly bodies above the horizon are measured by means of sextants or octants. The declinations of the principal heavenly bodies at various times, as well as the times at which they cross the meridian, are given in the Nautical Almanac.

Again, suppose that the altitude of the pole star be observed. From the figure it is seen that if the pole star were on the polar axis of the earth, then the altitude of the pole $P'XO$ would equal the latitude λ. The angular distance of the pole star from the polar axis of the earth, at various times, is given in the Nautical Almanac.

From the data for any star as given in the Nautical Almanac, together with the observed altitude at any known time not too remote from the time of crossing the meridian, the latitude of the observer can be computed.

72. Gyro-Horizons. — The latitude of a place is determined from the altitude of a celestial body. For the measurement of the altitude of a celestial body we must have a horizontal surface of reference. The horizon plane is the most convenient reference plane *when the horizon is visible*. On land we could use also the free surface of an unaccelerated liquid, or a plane normal to an unaccelerated plumb-bob. On a ship at sea neither of the two latter devices is available because each is subject to any force that accelerates the motion of the ship, such as the forces that produce rolling, pitching, or a change in either speed or course.

These forces produce less effect on the direction of the spin-axle of a gyro-pendulum that is normally vertical than upon the direction of either a liquid surface or a plumb-bob. A single impulse deflects the spin-axle in the direction of the force to a slight extent and it also produces a motion of precession having a definite period. If the period of precession of the spin-axle is great compared with the period of the accelerative forces producing rolling or pitching, the effect of the alternating impulses during a complete period of precession will be nearly neutralized. If the period of precession is not less than one hour, the deflection of the spin-axle of a gyro-pendulum from the vertical, produced by rolling or pitching of the ship, will be very small. This small error due to precession is eliminated almost completely by using the mean position of the spin-axle at instants separated by a half period of precession. For this procedure it is desirable that the period of precession be as small as possible. The required long period is secured by mounting the gyro so that the center of gravity of the precessing system is at a short distance below the point of support (Art. 68). The desired short interval between two instants separated by a half period requires that the center of gravity of the precessing system be at a great distance below the point of support.

The first gyro-horizon was made in the middle of the eighteenth century by an English instrument-maker named Serson. It consisted of a thick disk balanced at a point above its center of mass and capable of spinning on the upper end of a vertical pin. When the disk was maintained in rapid rotation, the spin-axis quickly assumed a practically vertical position and the upper polished surface of the disk became practically horizontal. It was proposed that this surface be used as a base from which to measure the altitude of heavenly bodies when the horizon is obscured by clouds or haze. The apparatus was tested on a British warship but, unfortunately, the ship went down and both the instrument and the inventor were lost. Since that time, gyroscopic horizons have received little attention till quite recently.

73. The Schuler Gyro-Horizon. — This instrument* consists of a gyro mounted in a spherical bowl that floats in a vessel of liquid contained in an outer bowl, Fig. 91. The inner bowl is

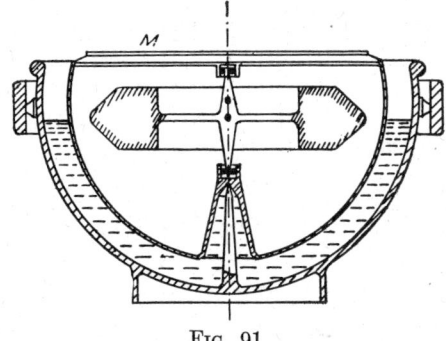

Fig. 91

kept centered by a pointed rod extending upward from the bottom of the outer bowl. The outer bowl is carried on gimbals. A plane mirror M is normal to the spin-axle. The center of gravity of the gyro with the inner bowl is at such a distance below the center of buoyancy that the period of precession of the spin-axle is about 85 minutes.

74. The Anschütz Gyro-Horizon. — This instrument† has a mirror M, Fig. 92, mounted on a gyro-casing supported on gimbals, the axes of which are at different levels above the center of gravity of the gyro and casing. Normally, the spin-axle is vertical. The housing carrying the gimbal axes can be turned in azimuth till the gimbal axis AA nearer the center of gravity of the gyro and casing is in line with the heavenly body under observation. As forces acting at the center of gravity and producing torques about the axis AA nearer the center of gravity have a short lever arm, the precession which they produce about the other gimbal axis BB is small. Consequently, the torque produces

* U. S. Patents. Schuler, No. 1480637, 1924; No. 1735058, 1929.
† U. S. Patent. Anschütz, No. 1141099, 1915.

only a small tilting of the spin-axis in the vertical plane of observation. A torque about the axis BB, further from the center of gravity, produces a precession of short period, as required for a quick succession of half-period observations.

The amplitude of oscillation of the spin-axle is damped by the friction of a mass of liquid moving back and forth in an annular trough D fastened to the gyro-casing concentric with the spin-axle.

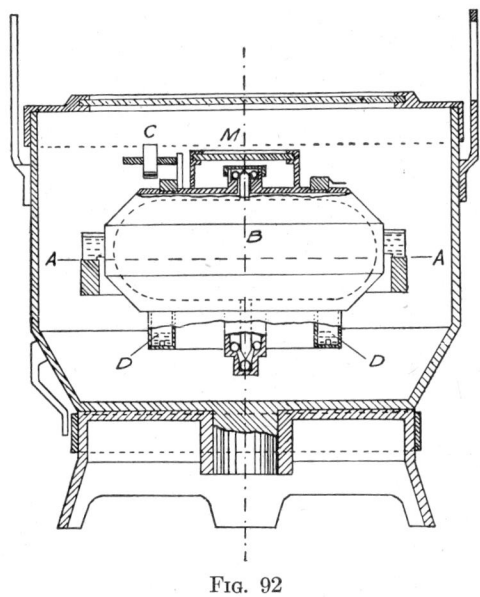

Fig. 92

Tilting of the spin-axle, due to the rotation of the earth, can be neutralized by a torque capable of producing an equal precession toward the east (Art. 41). The required precession is produced by the weight of a nut C that can be adjusted in position in azimuth and also in distance from the spin-axis. This nut must be either north or south of the spin-axle, depending on the direction of spin, and at a distance from the spin-axle depending upon the latitude. In setting the instrument for measuring the altitude of a celestial body, the instrument is turned till the gimbal axis nearer the center of gravity of the gyro and casing is in the vertical plane of the body under observation, the horizontal ring carrying the adjusting nut C is rotated till the nut is in the geographical meridian and on the proper side of the spin-axle, and the nut is moved horizontally to the scale position corresponding to the latitude of the place.

75. The Bonneau-LePrieur-Derrien Gyro-Sextant. — This is a light instrument designed for the use of aviators. It consists of a sextant to which is attached a gyro-horizon. Without reference to a visible horizon, altitudes can be measured with it to a precision within about fifteen minutes of angle. Attached to the sextant frame is a case in which is a top G, Fig. 93, capable of rapid rotation. The top is provided with a row of blades about the periphery. It is set into rotation and maintained at a high speed by means of either a few strokes of a bicycle pump or by gas escaping from one of the little steel capsules called "sparklets" filled with liquefied carbon dioxide and which are sometimes used to aerate beverages. The upper surface of the top is a plane mirror which maintains a nearly horizontal position while the top is spinning rapidly about a vertical axis NG.

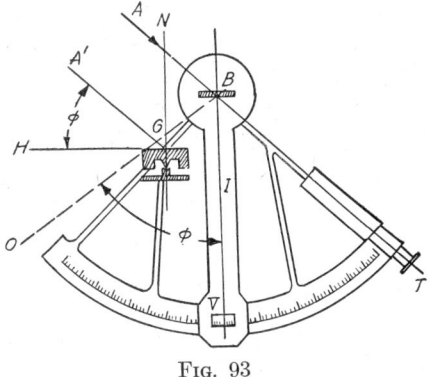

Fig. 93

Pivoted to the center B of the divided arc of the frame of the instrument is an index arm I which carries at one end a vernier scale V and at the other end a plane mirror B. The mirror is divided into two parts, one part being silvered and the other part being transparent. The surface of the mirror is in the plane of the axis of the pivot and is perpendicular to the axis of the index arm. The axis of the telescope is perpendicular to and intersects the axis of the pivot. The divisions of the scale are reckoned from a zero O on a line through the pivot perpendicular to the axis of the telescope.

When the telescope is pointed to a star, light from the star traverses the transparent part of the index mirror and forms an image in the focal plane of the telescope eye-piece. Light from the star is also incident on the horizontal gyro-mirror G. The index arm can be turned so that light after reflection from the gyro-mirror and the index mirror will form another image superposed on the first.

Since $A'G$ and AT are two parallel lines cut by GB, we have the angle

$$A'GB = GBT \quad \text{or} \quad \tfrac{1}{2}(A'GB) = \tfrac{1}{2}(GBT) = NGB$$

When the light after reflection from the two mirrors coincides with the axis of the telescope, the angle

$$GBV = VBT = \tfrac{1}{2} GBT$$

Whence, the angle $NGB = GBV$

Consequently, NG and BV are parallel, and the two mirrors are parallel. Since the spin-axis NG of the top is practically vertical, the two mirrors are practically horizontal.

From the construction of the divided circle, the line BO from the pivot to the zero line is perpendicular to AT and $A'G$. And since BV is perpendicular to the horizontal HG, it follows that the altitude of the star $A'GH = OBV$.

76. The Fleuriais Gyroscopic Octant. — This is an octant (or a sextant) to which is attached a top with a flat surface normal to the spin-axis and with the center of mass below the peg. The recent models carry on the flat surface, Fig. 94, a converging lens L and a piece of plane clear glass G perpendicular to the plane of the diagram.

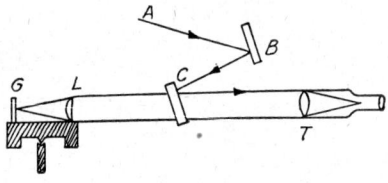

Fig. 94

One face of this glass is ruled with a series of parallel horizontal lines. The ruled face is in the focal plane of the lens. Light from the ruled lines, after traversing the lens L, the fixed half-silvered mirror C and the objective lens of the telescope T, forms an image of the rulings in the focal plane of the eye-lens. In the focal plane of the eye-lens there is a cross hair perpendicular to the plane of the frame of the instrument, that is, the cross hair is nearly horizontal. When the plane of the frame of the instrument and the axis of the top are vertical, light from the middle ruling traverses the axis of the telescope and the image of this ruling coincides with the cross hair. The axis of the telescope is now horizontal.

An eye of an observer at the eye-lens of the telescope perceives the image of the rulings every time light from the rulings proceeds along the axis of the telescope. The image is produced once during each revolution of the top. The impression of an image on the human retina persists for about one-tenth of a second after the cessation of the exciting cause, the duration depending upon the brightness of the light. Consequently, if the top be rotated at a sufficiently high speed, the image will appear to be continuous.

GYRO-HORIZONTALS AND GYRO-VERTICALS 121

The top is enclosed in an air-tight case K, Fig. 95, provided with a pair of windows in line with the axis of the telescope. When the instrument is used during the daytime the ruled glass plate is illumined by daylight entering the window farthest from the telescope. When used at night, the ruled plate is illumined by a tiny incandescent lamp operated by a pocket dry battery.

The required high spin-velocity is produced by a current of air blowing against a row of blades around the periphery of the top. A hand-operated exhaust pump is connected to a tube E and the case is evacuated to a pressure of from six to eight centimeters of mercury. The pressure is indicated by an aneroid barometer on

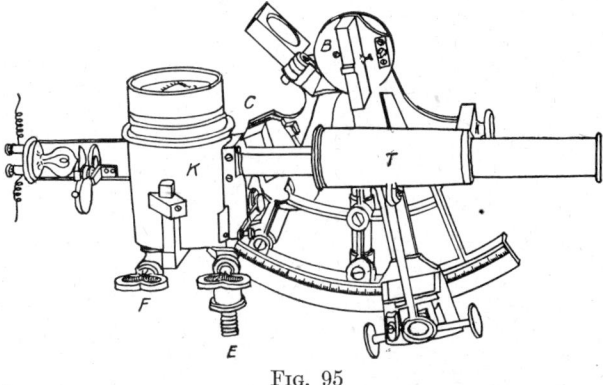

Fig. 95

top of the case. On now opening the stop-cock F, air rushes into the case and against the blades on the periphery of the top. The top quickly attains a speed of more than 7000 revolutions per minute. The stop-cock F is again closed and the pump operated till the pressure is again reduced to about six centimeters of mercury, the stop-cock E is closed and the pump detached. The top will continue to rotate in the vacuous space for several minutes without the spin-velocity diminishing as much as 25 per cent. This time is long enough to take the required observation.

If the spinning wheel rested on a sharp point, there would be a small angle between the spin-axis and the vertical that could be computed from (86). In the actual instrument, however, this angle is reduced to a negligible value by rounding off the peg that supports the spinning wheel (Art. 45).

77. The Sperry Roll and Pitch Recorder. — This instrument is used to record the angles of roll and of pitch of a ship in a seaway, together with the periods of the oscillations. It is a pendulous gyroscope* mounted so as to be capable of rotation about two perpendicular intersecting axes that normally are horizontal. The degree of pendulousness is such that the instrument has high dynamic stability together with a period of vibration that is many times greater than the period of either the roll or the pitch of any ship. Hence the direction of the spin-axle is practically unaffected by the oscillations of the ship. The vibrations of the instrument

FIG. 96

are damped by the motion of a mass of liquid contained in an annular passage concentric with the spin-axis. The passage is constricted so that the flow of liquid back and forth will be sufficiently out of phase with the oscillations of the gyro to produce the desired damping effect.

Figure 96 shows the instrument as mounted on a ship with the outer gimbal axis athwartship and the inner one in the fore-and-aft direction. The pen nearest the divided scale records the angle of roll on a strip of paper that is moved parallel to the keel of the ship at a constant rate by means of a clock mounted on the bed

* U. S. Patent. Sperry, No. 1399032, 1921.

plate. The pen to the left records angles of pitch and the pen to the right indicates time intervals.

The roll-recording pen is held by a horizontal arm fastened to a vertical counterbalanced circular loop capable of rotation about pivots in line with the fore-and-aft gimbal axis of the instrument. The concave side of the upper part of this loop is deeply grooved. Into this groove projects a rod fastened to the top of the gyro-case and in line with the spin-axis. When the ship rolls, the spin-axle and the loop remain in the vertical plane while the paper carrier moves athwartship under the pen. If at the same time the ship pitches, the groove in the loop permits motion of the loop relative to the paper carrier without affecting the movement of the paper relative to the roll-recording pen.

The pitch-recording pen is supported by a parallel motion device that permits motion of the pen only in the athwartship direction. It carries two horizontal guides that for part of their length are inclined at about 45 degrees to the fore-and-aft axis of the ship. In the space between the two guides is the upper end of a vertical rod, the lower end of which is fastened to the outer gimbal ring. When the ship pitches, the parallel motion device moves relative to the gyro spin-axle along the fore-and-aft axis of the ship. In so doing, the guides are pushed in the athwartship direction by the rod extending upward from the outer gimbal ring. Thus the attached pen traces a curve that indicates the angle of pitch with respect to time.

78. The Sperry Automatic Airplane Pilot. — This apparatus maintains an airplane on any predetermined straight course as long as desired. It is especially useful in " blind flying " through opaque clouds. The aviator may leave the cockpit for any purpose, may even walk out to the end of a wing, with confidence that the machine will continue in its course on a horizontal keel. If he were to faint for a brief period the automatic pilot would keep the machine in a straight course on a horizontal keel. He can disconnect the automatic steering device instantly whenever he decides to take control.

The automatic pilot comprises two universally mounted gyros, one with spin-axle vertical and the other with spin-axle horizontal. The first keeps the airplane horizontal by controlling the position of horizontal rudders. The second maintains the course of the airplane in azimuth by controlling the position of vertical rudders. The power required to spin the gyros is supplied by the electric system of the airplane, while the power to operate the rudders is supplied by air turbines which rotate when the airplane is in motion.

The directional gyro, that is the gyro which by operating vertical rudders keeps the machine on its straight course, is shown in Figs. 97 and 98. If the airplane turns in azimuth, the gyro-frame turns relative to the spin-axle; two contactor plates attached to the frame slide along a trolley not shown in the figures; an electrically operated clutch connects one of the air turbines to the vertical rudder thereby automatically turning the airplane back into the course.

Fig. 97

The frictional forces acting at the bearings of the gyro are so small that the spin-axle will remain fixed in space for several minutes, but after a time the spin-axle will be appreciably out of its original direction. So long as the airplane is on a straight course the degree of constancy in direction of the spin-axle can be checked by comparison with a spirit level and magnetic compass. When the spin-axle does become tilted from the horizontal, an air-blast escaping from an orifice in the gyro-casing O, Fig. 97, produces a torque about a vertical axis, which torque automatically brings the spin-axle back into a horizontal plane. When the spin-axle is horizontal, the orifice is covered by a shutter S and the above-described torque is not produced.

When the aviator observes that the spin-axle has moved in

azimuth from the original setting, he presses an electric button, thereby energizing one of a pair of curved solenoids A, A', Fig. 98, attached to the gyro-frame. This causes a curved iron core attached to the gyro-casing to be pulled toward the middle of the solenoid, thereby developing a torque on the casing, about a horizontal axis, in the proper direction to cause the spin-axle to precess about a vertical axis back into the original direction. This operation moves the contactor attached to the gyro-casing, relative to the trolley, thereby causing one of the air turbines to turn a vertical rudder in the proper direction to bring the keel of

FIG. 98

the airplane parallel to the new position of the spin-axle, that is, to bring it into the original course.

There are two electric buttons, curved solenoids and curved cores, one system to cause a turning of the airplane in the clockwise direction and another to cause a turning in the counter-clockwise direction.

The second gyro operates in a manner similar to the directional gyro, above described, except that this gyro with vertical spin-axle operates horizontal rudders so as to maintain the axis of the airplane in a horizontal position.

79. The Sperry-Carter Track Recorder. — This is a group of instruments[*] that while carried on a test car moving at train speed

[*] Railway Engineering and Maintenance, Vol. 23, p. 316.

will make a chart of the magnitudes and positions of all irregularities of the track. On the paper are drawn curves coördinating distance and time, distance and differences in the elevation of the two rails under the car as well as the positions and magnitudes of rail spreads and rail depressions.

Differences in the elevation of the two rails are recorded by means of a gyroscopic apparatus mounted on a table supported by one of the car axles, Fig. 99. The function of the gyro is to furnish, at any instant, a vertical plane parallel to the rails, with reference to which can be measured any elevation of one rail above the other.

Fig. 99

The gyro is mounted with the spin-axle normally parallel to the car axles. The gyro-casing is supported in a ring R, Fig. 100, which normally is vertical. The gyro-casing is capable of turning about a vertical axis, and the ring is capable of turning about a horizontal axis parallel to the rails. The center of gravity of the gyro and of all parts attached to it coincides with the intersection of these three axes. Hence, a change in either the direction or magnitude of the velocity of the car produces no torque on the gyroscopic system and consequently no change in the direction of the spin-axis in space.

When the car axle tilts from the horizontal, the frame of the gyroscope tilts with respect to the spin-axle, thereby causing a paper record roll to move under the pen p perpendicularly to the plane of the diagram.

For accurate indications, the indicator post I must be maintained in a vertical plane perpendicular to the normal position of the car axle. The indicator post is maintained in this vertical plane by means of a pendulum P and a pair of current-carrying solenoids SS. These solenoids are fastened to the ring R. Projecting into each solenoid is one end of a soft iron core having the other end fastened to the gyro-casing. When either of the solenoids is traversed by a current, the far end of its core is drawn inside the solenoid thereby producing a torque on the gyro-casing about a vertical axis.

If, for any reason, the indicator post becomes deflected out of a vertical plane perpendicular to the car axle, that is, out of the plane of the diagram, Fig. 100, the contact arm C moves relative to a trolley wheel T on the upper end of the pendulum P. An electric circuit is thereby completed through one of the solenoids S, a torque acts on the gyro-casing about the vertical axis, and the ring R precesses about a horizontal axis till the indicator rod is parallel to the pendulum.

Since the plane of vibration of the pendulum is perpendicular to the direction of motion of the car, no change in velocity of the

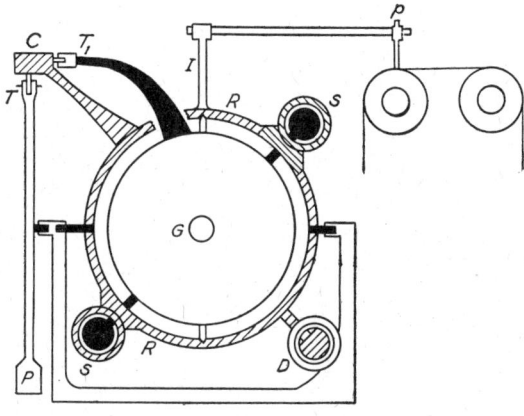

Fig. 100

car while it is moving on a straight track will deflect the pendulum from the vertical. When, however, the car is moving around a curve, the pendulum bob will retreat from the center of the curve, thereby deflecting the pendulum from the vertical. Consequently, during the time that the car is going around a curve, the electric circuit controlled by the pendulum must be kept open. An arrangement of contacts automatically keeps the pendulum circuit open while the car is going around a curve. During this time, the direction of the spin-axle is fixed in space because the gyro is mounted so as to be free of any torque due to any change in either the direction or magnitude of the velocity of the car.

While the car is going around a curve, the spin-axle of the gyroscope tends to turn *relative to the car* about a vertical axis. This is due to the fixity *in space* of the spin-axle. Again, the friction of the pen on the paper produces a torque about the horizontal axis of rotation of the ring. This torque also tends to produce precession of the spin-axle about the vertical axis.

The spin-axle should be kept as nearly parallel to the car axles as possible because this is the position in which the gyroscope is least affected by pen friction. For this reason, just as soon as the spin-axle is deflected from parallelism to the car axle by as much as two degrees, a torque is applied to the ring R about the horizontal axis and of sufficient magnitude to prevent the turning of the spin-axle about the vertical that otherwise would occur. The required torque is produced by a pull on one or the other of two soft iron cores, each having one end attached to the ring R and the other end within one of two solenoids D fastened to the frame of the apparatus. The circuit through one solenoid or the other is made or broken as the trolley T_1 moves from one contact plate across a strip of insulation to another contact plate on the contact arm C.

By adding another control pendulum capable of turning about a horizontal axis perpendicular to the one used in this apparatus, an apparatus could be produced that would furnish a horizontal plane which could be used wherever an artificial horizon is required.

80. Directed Gun-Fire Control. — It is a complex problem to make the observations and computations from which a gun on a pitching and rolling ship may be directed so that the projectile may strike another ship that is moving with respect to the first ship and that is invisible to the gunners. The target must be visible to an observer who is in communication with computers, and the latter must be able to transmit orders to the gunners. The observer may be on a mast above the smoke of battle, or he may be in an airplane at such a height that he can see the target which may be below the horizon of the gunners.

The observer notes the direction of the line from the target to the mother ship, the range, that is the distance between the target and the ship, and he estimates the course and speed of the target. These data are communicated to the computers. They know the course and speed of their own ship, the drift of the projectile from each gun (Art. 48), as well as the approximate deflection produced by the wind. There is an additional deflection when the axis of the gun trunnions is not horizontal. If the right end of the trunnion-axis dips, the projectile will be deflected to the right of the line of sight by an amount depending on the angle of dip. All angles in azimuth must be measured from a base line which is fixed relative to the earth. This base line is indicated by a **gyro-**

compass of the highest degree of precision known to the art. Such an instrument is called a " gun-fire control compass," whereas an instrument of a precision sufficient for navigational purposes, but not for gun-fire control, is called a " navigational compass."

These data are set up on the dials of a computing machine called a " range clock " or " range and bearing keeper." This instrument then indicates automatically and continuously what the range and the bearing of the target will be at any assigned subsequent instant so long as the data remain unchanged.*

The appropriate angles of elevation and of train are transmitted to indicating instruments situated at the guns. The gunners keep the gun pointed and trained in the directions given by the indicators. When the ship has rolled to the proper angle the gun is fired, either automatically or manually, as desired. An observer called a " spotter," stationed either on a mast or in an airplane, notes where the projectile strikes and informs the computers. The angles transmitted to the gunners are revised till the target is hit.

As the various gun-fire control computing machines do not depend on gyro-dynamics, they need not be described here. Various sighting instruments and automatic firing mechanisms, however, are stabilized by gyro-pendulums.

81. Gun-Fire Directorscopes. — For a given angle of gun elevation with respect to the deck of a ship, there is but one angle of roll of the ship at which a projectile fired from the gun will strike the target. The optical system of a sighting telescope on a ship may be stabilized by a gyro-pendulum so that the image of a distant target will remain stationary in the field of view however the ship may roll. A co-acting firing mechanism may be set so that when the ship has rolled to an assigned angle with respect to the horizontal, one or more guns will be discharged. If the guns were correctly loaded, correctly pointed with respect to the deck of the ship, and correctly trained with respect to the meridian, the projectiles will hit the target. An apparatus consisting of one or more sighting instruments combined with a co-acting firing mechanism, designed to discharge, in the proper direction and at the proper instant in the roll of the ship, the guns controlled by the apparatus, is called a gun-fire directorscope.

* U. S. Patents. Ford, No. 1370204, 1921; No. 1450585, 1923, No. 1472590, 1923; No. 1484823, 1924. Meitner, No. 1455799, 1923. E. Sperry, No. 1296439, 1919; No. 1356505, 1920; No. 1755340, 1930.

Angles in azimuth can be laid off relative to the N–S line of a gyro-compass, and angles in elevation with respect to a gyro-vertical or gyro-horizontal forming part of the apparatus. An electric circuit through the gun, gyro-compass and firing mechanism may be closed automatically at the instant the gun axis makes the proper angle to the N–S line of the gyro-compass, and the proper angle of elevation relative to a gyro-pendulum forming part of the firing mechanism.*

* U. S. Patents. Schneider, No. 1507209, 1924; Radford, No. 1531132, 1925; Ford, No. 1597031, 1926; Crouse, No. 1689327, 1928.

CHAPTER IV

GYROSCOPIC ANTI-ROLL DEVICES FOR SHIPS

§1. *The Oscillation of a Ship in a Seaway*

82. The Rolling of a Ship Due to Waves. — From Archimedes' Principle, a body either wholly or partially immersed in a fluid is buoyed up by a force equal to the weight of the fluid displaced. The weight of a body acts at a point called the center of gravity of the body. The buoyant force acts at a point called the center of buoyancy. The center of buoyancy is at the center of mass of the fluid displaced.

If the center of gravity of a completely immersed body is below the center of buoyancy, the body is statically stable, that is if this body be tilted and then released it will recover its former position. If the center of gravity of a completely immersed body is above the center of buoyancy, the body is unstable; that is if this body be tilted and then released, it will not recover its former position but will turn over till the center of gravity is below the center of buoyancy.

A floating body, however, may be statically stable when the center of gravity is above the center of buoyancy. Figure 101 represents a ship standing upright in still water with the center of gravity of the vessel at G and the center of buoyancy at B. Suppose that a wave moves under the ship from the left to the right. At the instant represented in Fig. 102, the center of buoyancy has moved to the left of the line of action of the weight F_g. The weight F_g and the buoyant force F_b now constitute a couple having a lever arm x. This couple causes the ship to "heel" or roll, in the clockwise direction. If the water surface is again level when the ship has rolled into the position shown in Fig. 103, the center of buoyancy is to the right of the line of action of the weight. Now the couple tends to roll the ship back into the upright position. The advancing wave emerging from under the ship adds a couple in the same direction.

The angular amplitude of vibration of a ship produced by a single wave is always small. If, however, a series of waves pass under the ship at regular intervals, and the period of the oncoming waves is nearly equal to the period of roll of the ship, then the angular amplitude of roll quickly builds up to a considerable value. The building up of a large amplitude of oscillation by the cumulative effect of a periodic disturbance having nearly the same period as that of the vibrating body is called resonance (Art. 26). The large angle of roll, so common with ships at sea, occurs when the period of the oncoming waves is nearly equal to a natural period of oscillation of the ship. The period at which the oncoming waves meet the boat depends on how fast they travel, the distance from one

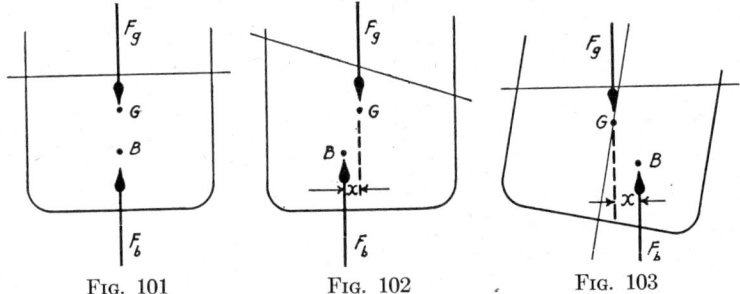

FIG. 101   FIG. 102   FIG. 103

wave to the next, the course of the boat across the waves, and the speed of the boat.

83. The Pitching of a Ship Due to Waves. — If either the bow or the stern of a ship with a single propeller be suddenly raised or lowered, the precession thereby produced will deflect the ship's course either to the right or to the left. This "yawing" or "nosing" from side to side results in the pushing aside of a larger mass of water by the moving ship and a consequent loss of power and speed. Pitching increases the difficulty of maintaining the ship's course and increases the fuel consumption. The yawing produced by pitching can be neutralized by the use of two similar propellers and shafts rotating in opposite directions. This device, however, does not diminish the up-and-down pitching motion with the accompanying discomfort to passengers, danger of injury to freight, and strains within the frame of the ship, nor does it reduce the natural yaw due to a wave not striking the bow and stern at the same time.

84. The Metacentric Height. — When a vessel floats upright, the center of gravity G and the center of buoyancy B are on the same vertical central line AA', Fig. 104. When the vessel is heeled over, the center of buoyancy is displaced to one side of the central line. So long as the heeling does not exceed about 10 degrees, the line of action of the buoyant force intersects the central line AA' near the same point M. That point in a floating body slightly displaced from equilibrium, through which the resultant upward force of the displaced liquid intersects the vertical through the center of buoyancy when the body is not displaced, is called the *metacenter* of the body. If a ship be slightly tilted from the equilibrium position, the intersection M, of the line of action of the force of buoyancy and the line perpendicular to the decks of the ship through the center of gravity, coincides with the metacenter of the ship. The metacenter for rolling does not coincide usually with the metacenter for pitching.

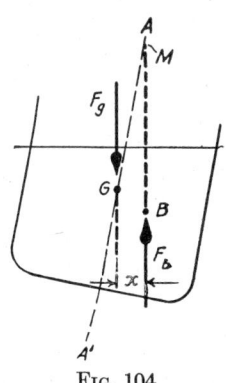

Fig. 104

If the metacenter is above the center of gravity, the vessel is stable; if it is below the center of gravity, the vessel is unstable; if it coincides with the center of gravity, the vessel is statically neutral. The distance MG from the metacenter to the center of gravity is called the *metacentric height* of the vessel. The degree of stability of a vessel depends upon the metacentric height.

A vessel with a metacentric height that is large in comparison with the size of the vessel has a large righting moment, a short period of roll, is very stable, but will roll quickly through large angles and will change its direction of roll with a jerk. Such a vessel is difficult to steer, requires an excessive amount of fuel for a given speed, and is uncomfortable for crew and passengers. Many cargo steamers carrying coal or ore are uncomfortable and hard to handle because of their great metacentric height. If this cargo could be distributed so as to raise the center of gravity of the vessel the severity of these troubles would be diminished.

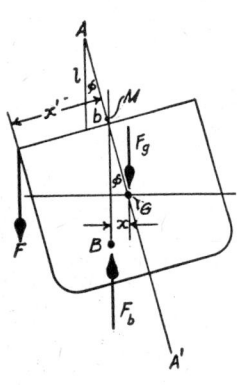

Fig. 105

85. The Experimental Determination of the Metacentric Height. — The metacentric height of a vessel of known tonnage displacement can be determined from an observation of the tilting produced by shifting a known weight of crew or ballast for a known distance across the deck. In Fig. 105, a vessel of known weight F_g has been tilted from the upright position through the small angle ϕ by moving a load F across the deck through a distance x'. The center of gravity of the vessel with cargo is at G, the center of buoyancy at B, and the metacenter at M. The small angle through which the ship has been tilted is measured by means of a plumb-bob of length l hung at A. Represent by the symbol b the distance along the deck that the plumb-bob has moved when the vessel is tilted. Represent the metacentric height MG by the symbol H.

Now the moment of the tilting force equals the moment of the righting force. Taking moments about the metacenter and assuming that the angle of tilting is so small that x' and b are approximately equal to their horizontal projections,

$$Fx' = F_g x = F_g H \sin \phi = F_g H \frac{b}{l}$$

Whence, the metacentric height

$$H = \frac{Fx'l}{F_g b} \tag{97}$$

86. The Period of the Rolling Motion of a Ship. — By the period is meant the time of a complete back-and-forth oscillation. When a vessel has been rolled through an angle ϕ there is acting upon the vessel a righting torque, Fig. 105, having the value

$$L = F_g x = F_g H \sin \phi$$

where H is the metacentric height.

This equation shows that when the angle of roll ϕ is so small that $\sin \phi$ may be replaced by ϕ radians, then the restoring torque is directly proportional to ϕ. This is the law of simple harmonic motion of rotation (Art. 20). Consequently, if a vessel be rolled through a small angle from its equilibrium position, and if it be unacted upon by any torque except the righting torque, it will roll back and forth with a periodic motion that is approximately simple harmonic motion of rotation. On substituting in (37) the value of the torque $L = F_g H \phi$, we find that the period of such a simple harmonic motion of rotation is

$$T = 2\pi \sqrt{-\frac{K\phi}{L}} = 2\pi \sqrt{-\frac{K}{F_g H}} \tag{98}$$

where K is the moment of inertia of the vessel relative to the axis about which the rolling occurs. The negative sign indicates that φ is measured in the direction opposite to L. Wind and frictional forces, and the forces due to the impacts of succeeding waves, produce such additional torques that the rolling motion is not simple harmonic and the value of the period cannot be computed with precision.

If, however, the rolling of a ship be assumed to be simple harmonic, then the H in (96) represents the metacentric height. In this case, if a ship carries a gyroscope of great angular momentum and with the spin-axle and the precession axis perpendicular to the axis of roll, then the period of roll is altered as it would be by a lengthening of H to $\left(H + \dfrac{h_s^2}{mgK_c}\right)$. Equation (98) shows that an increase of the metacentric height of a ship produces a decrease in the period of roll. Consequently, the rolling of a ship can be caused to be quicker by mounting on the ship a large gyro capable of spinning and precessing as above indicated.

87. Methods of Diminishing the Amplitude of Roll. — Roll increases stresses in the ship's structure and engines; it increases the effective area of cross-section of ship that must be pushed through the water, and consequently the fuel consumption; it decreases speed; it decreases the comfort of passengers and crew; it decreases the accuracy, range and rapidity of fire from naval vessels. These are ample reasons for the serious efforts made to diminish or suppress roll. Anti-roll devices are commonly called " ship stabilizers " because by their use the tilting of the ship from the equilibrium position by an applied torque is diminished.

The most obvious device is to attach long planks lengthwise on the outside of the hull below the water line. These so-called " bilge keels " decrease the roll but they also decrease the speed of the ship. To avoid the excessive friction through the water when the stabilizing effect of the bilge keels is not needed, it has been proposed to use fins that can be moved in and out of longitudinal slits through the hull of the ship.* When the ship is rolling, the fins would be protruded; when the ship is not rolling the fins would be withdrawn.

Frahm's anti-roll tanks (Art. 28) have been used to a considerable extent, although their mass and the difficulty in adjusting the size of the connecting passage to proper operation, have limited the use of the Frahm system.

* U. S. Patents. Thompson and Schein, No. 1475460, 1923; Motora, No. 1533328, 1925; Kéféli, No. 1751278, 1930.

Several schemes have been devised, and some of them developed as far as the initial experimental stage, in which the property of the axle of a spinning gyro-wheel of maintaining its direction in space has been used to control an engine which in turn would move large masses to the higher side of a rolling ship. The magnitudes of the masses that would need to be moved has caused these suggestions to be neglected.

At the present time gyroscopic anti-roll devices are in successful use. They are much used on yachts and also to some extent on large ships. The action of these devices is based on the fact, proven in Art. 25, that if a body oscillating with a periodic angular motion be acted upon by a torque of the same period, and in the opposite phase to the velocity of the oscillating body, then the system producing the periodic torque will absorb energy from the oscillating body.

§2. *The Inactive Type of Gyro Ship Stabilizer*

88. The Effect on the Motion of a Swinging Pendulum Produced by an Attached Gyroscope: (a) When the Precession of the Gyro-Axle is Opposed by a Frictional Torque. *Experiment.* — The pendulum shown in Fig. 106 carries a gyro-wheel mounted so as to have two degrees of rotational freedom. The gyro-wheel spins about a nearly vertical axis through the center of mass of the wheel and attached frame, and it is capable of rotation about a perpendicular axis AA' in the plane of vibration of the pendulum. For this experiment, the wheel and its supporting frame are lowered till the latter axis is above the center of gravity of the gyro-wheel and its supporting frame. The gyro-frame is prevented from rotating too far out of the plane of vibration of the pendulum by means of stops. A brake B is provided by means of which the precession of the gyro-axle can be opposed by a torque of any desired magnitude.

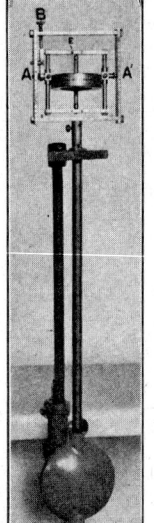

Fig. 106

Release the brake. Set the pendulum into oscillation while the gyro-wheel is spinning. Observe that there is negligible damping of the amplitude of oscillation of the pendulum when the precession is unopposed by the friction brake.

Release the thumbscrew that clamps the gyroscope to the pendulum. Set the pendulum swinging while the wheel is spinning.

THE INACTIVE TYPE OF GYRO SHIP STABILIZER

Observe that the gyroscope twists back and forth about the axis of the pendulum as the pendulum oscillates back and forth.

The angular acceleration of the motion of the pendulum is accompanied by a torque on the gyro-frame about an axis parallel to the knife-edge. This torque is maximum when the pendulum is at the end of its swing and is zero when the pendulum is at the

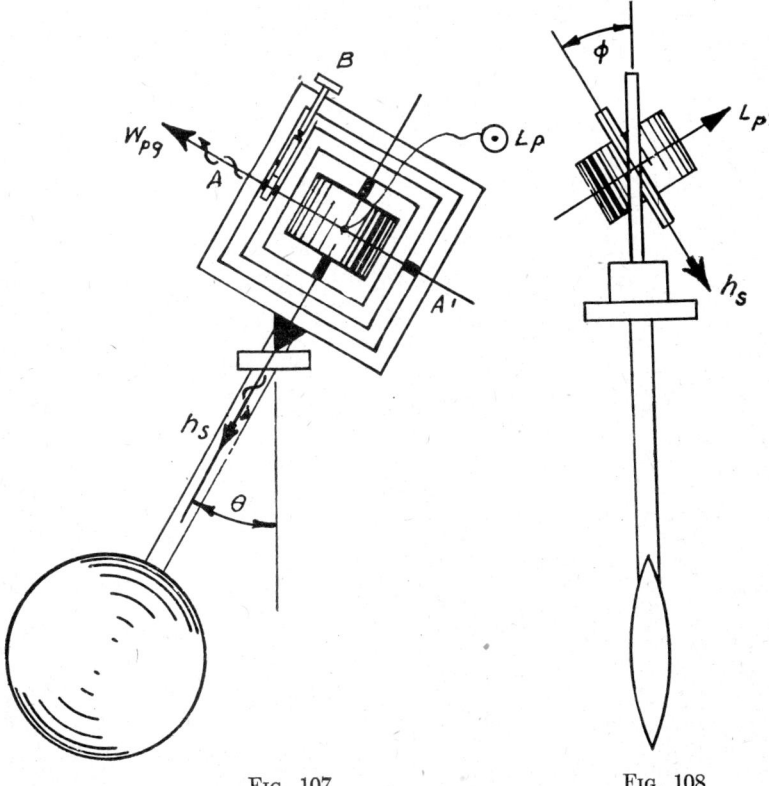

Fig. 107   Fig. 108

middle of its swing. It produces a precession of the gyro-axle about an axis AA', Fig. 107, perpendicular to the knife-edge. This precession produces a torque L_p on the pendulum about an axis perpendicular to the spin-axle and to the axis of precession. At any instant the magnitude of this torque is $L_p = h_s w_{pg}$, where h_s and w_{pg} represent the instantaneous values of the angular momentum of the gyro-wheel about the spin-axis, and the angular velocity of precession, respectively. Hence, when the gyro-axle is passing through its equilibrium position, the gyroscopic torque acting on the pendulum is maximum and consequently the deflecting torque on the pendulum is maximum.

The swinging of the pendulum and the precession of the gyro-axle have a common period. When the velocity of precession is maximum, the angle of precession ϕ is zero, Figs. 107 and 108. Consequently, when the deflection θ of the pendulum from the vertical is maximum, ϕ is zero. The variation of θ with respect to time is represented by the curve marked θ in Fig. 109. The variation of ϕ with respect to time is represented to a fair degree of accuracy by the curve marked ϕ.

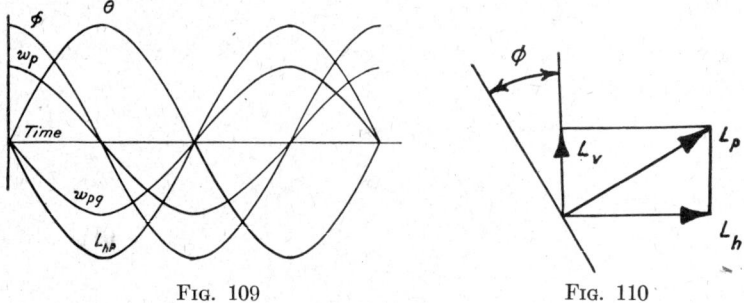

Fig. 109　　　　　Fig. 110

The angular velocity of the pendulum due to gravity is zero when the inclination of the pendulum to the vertical is maximum, and the velocity is maximum when the inclination is zero. The relation between the angular velocity of the pendulum and time is represented approximately by the curve marked w_p.

At the instant when the pendulum is at the end of a swing, θ is maximum, the gyro-axle is passing through its mid-position ($\phi = 0$), the speed of precession w_{pg} is maximum, and the torque L_P acting on the pendulum due to the attached gyroscope is maximum.

When the angle between the gyro-axle and the vertical is ϕ, Fig. 108, there is a vertical component of the torque acting on the pendulum due to the precession of the attached gyroscope of the value, Fig. 110.

$$L_v = L_P \sin \phi = h_s w_{pg} \sin \phi$$

This vertical component tends to twist the pendulum about the axis of the pendulum, as was observed in the experiment. The component about a horizontal axis parallel to the knife-edge is

$$L_{hP} = L_p \cos \phi = h_s w_{pg} \cos \phi$$

The variation of L_{hP} with respect to time is represented with a certain degree of precision by the curve marked L_{hP}, Fig. 109.

The power imparted to the pendulum by the gyroscope at any instant equals the product of the torque acting on the pendulum at that instant due to the gyroscope L_{hP} and the angular velocity of the pendulum, w_p. The average value of the power during one complete vibration is given by a curve of products of the instantaneous values of L_{hP} and w_p. It will be observed that the curves of L_{hP} and w_p are in quadrature, that is, they differ in phase by 90 degrees or a quarter of a period. Now the average value of the product of two sine curves of the same frequency is zero when the curves are in quadrature (Art. 25). That is, if the relation between L_{hP} and time, and that between w_p and time, were accurately represented by sine curves in quadrature, then the average power applied to the pendulum throughout one vibration by the precessing gyroscope would be zero. Consequently, under the conditions specified, the vibrations of the pendulum are undamped.

If, however, the phase difference between the L_{hP} curve and the w_p curve were greater than 90 degrees, then the power imparted to the pendulum by the precessing gyroscope would be negative, that is, energy would be abstracted from the pendulum. The more nearly the two curves are to being in opposite phase the greater will be the damping. They may be brought more nearly into this condition by retarding the L_{hP} curve relative to the w_p curve. This can be done by opposing the motion of precession w_{pg} by a friction brake. Friction at the bearings AA', Fig. 107, constitutes a torque about that axis in the direction opposite the precession. This torque produces a precession of the inner gyro-frame about an axis perpendicular to the plane of the diagram which opposes the velocity of the pendulum and increases the phase difference between L_{hP} and w_p. The power absorbed from the pendulum is now no longer zero. The vibrations of the pendulum are damped. The amplitude of vibration of the pendulum would not be damped if there were no opposition to the precession of the gyro.

The method here indicated for bringing variations of the precessional velocities of the gyro-axle of a spinning gyroscope more nearly into opposite phase with the variations of the torques acting on an oscillating body that carries the gyroscope is the basis of the action of the ship stabilizers of the inactive type designed by Schlick, Fieux, and others.

89. The Effect of a Spinning Gyroscope on the Rolling of a Ship.

— Consider a ship on which is mounted a gyro-wheel spinning about a vertical axis. Let the frame supporting the gyro-wheel be capable of rotating about an athwartship axis. When the ship rolls, the gyro-axle precesses about this axis AA', Fig. 111. The effect of the precession of the gyro-axle on the rolling motion of the ship is the same as that of the gyroscope on the motion of the pendulum considered in the preceding Article. If the gyro-axle precesses freely, there is zero damping effect; if the preces-

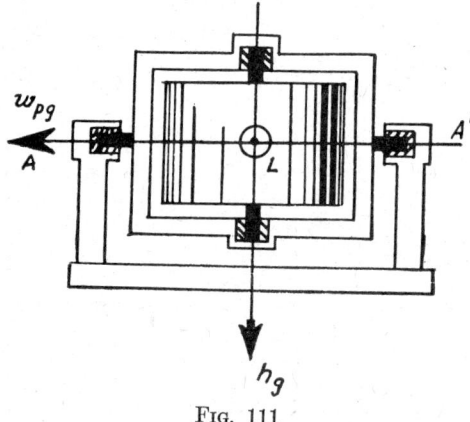

Fig. 111

sion is opposed by a moderate frictional torque, damping of the amplitude of roll is produced. The amount of damping would be maximum if the torque opposing precession at each instant were proportional to the speed of precession at that instant. Steering is not affected by a gyro-wheel with vertical spin-axis. The period of roll of the ship is decreased, (Art. 86).

A gyro-wheel spinning about an athwartship axis and capable of rotating about an axis AA' perpendicular to the decks of the ship, Fig. 112, will precess back and forth about this axis as the ship rolls from side to side. As in the case of the gyro with vertical spin-axle, the amplitude of roll of the ship will be damped if the precessional motion be opposed by a frictional torque.

Suppose the direction of spin is that represented by the line h_s, Fig. 113. Then while the ship is rolling to starboard with angular velocity represented by the symbol w, the gyro-frame and ship will be subjected to a torque about an axis perpendicular to h_s and w in the direction represented by the line L, (Art. 36). Owing to this torque the ship steers to starboard. Similarly a roll to

THE INACTIVE TYPE OF GYRO SHIP STABILIZER   141

port causes the ship to steer to port.* For this reason a single gyro-wheel with spin-axle horizontal cannot be used to stabilize a rolling or pitching ship.

This yawing can be prevented by using two gyro-wheels G_1 and G_2 that are spinning in opposite directions about horizontal

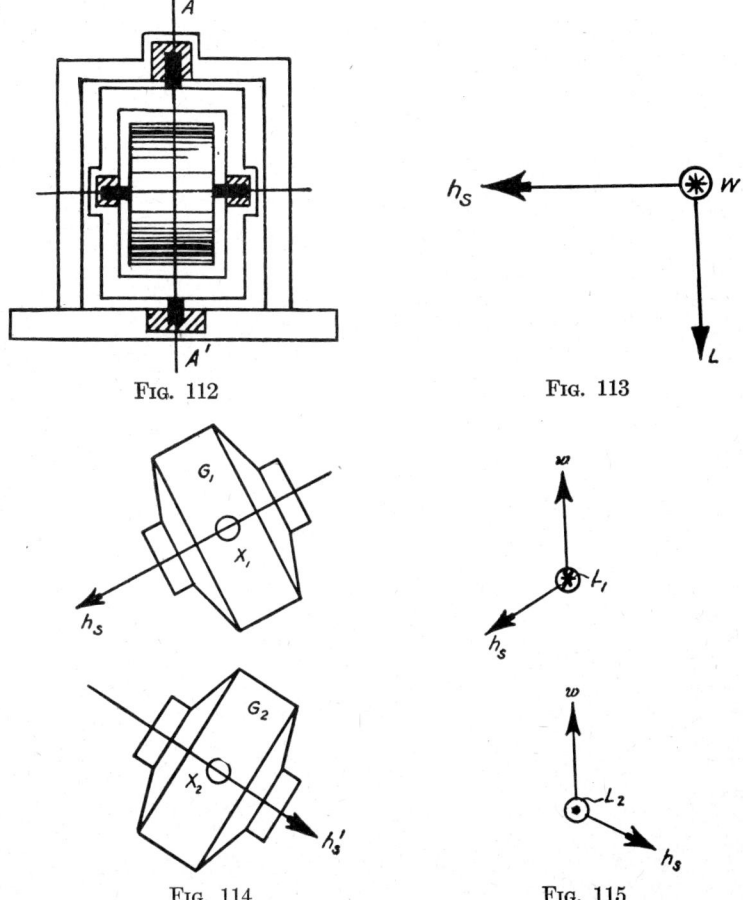

Fig. 112   Fig. 113

Fig. 114   Fig. 115

athwartship axes, Fig. 114. These gyro-wheels are capable of rotation about vertical axes x_1 and x_2, respectively. When the ship rolls, the two gyros precess in opposite directions about these axes. As the spin velocities are also in opposite directions, the torques developed by a roll of the ship are in the same direction.

* Suyehiro, "Yawing of Ships," Trans. Inst. Nav. Arch., 1920, pp. 93–101.

If the precessional motion of each is opposed by a moderate frictional torque, the amplitude of ship's roll will be damped.

In Fig. 115, the velocity of roll at some instant is represented by w. From Art. 36, the torque acting on the frame of G_1 about a vertical axis is in the clockwise direction as represented by the symbol L_1, and that acting on the frame of G_2 is in the counter-clockwise direction as represented by the symbol L_2. As these torques are equal and oppositely directed, the pair of precessing gyro-wheels produces zero effect on the steering of the ship.

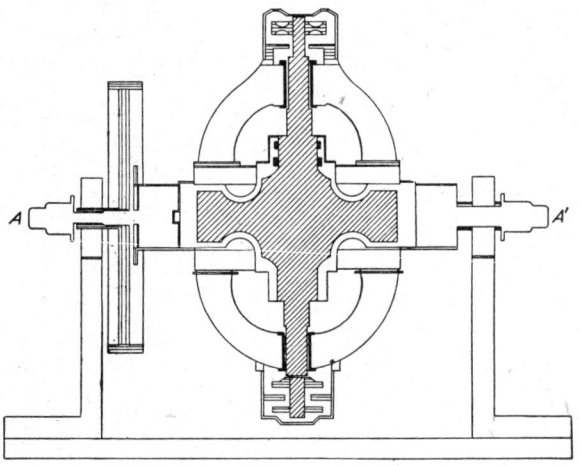

Fig. 116

90. The Schlick Ship Stabilizer. — This consists of a gyro-wheel of great moment of inertia* spinning about a vertical axis and capable of precessing about a horizontal athwartship axis AA', Fig. 116. The precession axis is above the center of gravity of the gyro-wheel and casing. The period of the apparatus when swinging as a gyro-pendulum is made approximately the same as the period of roll of the ship in a calm sea. The inclination of the gyro-axle from the equilibrium position is maximum when the rolling ship is in its equilibrium position, and it is zero when the ship is at the end of a roll. When there is no opposition to precessing, there is a phase difference of 90 degrees between the precessional velocity of the undamped gyro-axle and the gyroscopic torque acting on the ship (Art. 89). In this case there is zero

* U. S. Patent. Schlick, No. 769493, 1904.

absorption of energy from the ship's roll and consequently no damping of the amplitude of oscillation.

To produce an absorption of energy of the ship's roll, that is, a damping of the amplitude of roll, the phase difference between the precessional velocity and the torque acting on the ship due to the precessing gyro-wheel must be more than 90 degrees. In the Schlick ship stabilizer, the phase difference is increased by means of a brake* applied to the precession axle. A considerable part of the energy of the waves is used in producing precession of the gyro-axle and then is transformed into heat at the brakes.

Professor A. Föppel has shown† that if precession of the gyro-axle be opposed by the proper frictional torque, and if the moment of inertia of the gyro-wheel about the spin-axis has the value

$$K_s = \Phi \frac{\sqrt{F_g H K}}{5 \, w_s} \tag{99}$$

then, during the time of one roll, a Schlick ship stabilizer will reduce the velocity of roll through the equilibrium position to 0.283 of the value it would have when the gyro-wheel was not operating. In this equation, Φ is the maximum amplitude of roll without stabilizer, F_g is the weight of the ship, H is the metacentric height, K is the moment of inertia of the ship about the axis of roll, and w_s is the spin velocity of the gyro-wheel.

An essential part of this device for diminishing ship's roll is the brake. The brake torque at each instant should be proportional to the precessional velocity. Schlick has used band brakes and hydraulic brakes. When the brakes are not applied, the zero precessional velocity of the gyro, w_{pg}, and the zero angle of roll of the ship, θ, occur at the same instant, Fig. 109. Even with moderate damping they occur at near the same instant. A ship must roll through an appreciable angle before the precessional velocity of a Schlick gyro-stabilizer, and the torque thereby produced, become of sufficient magnitude to produce an effective opposition to the roll.

* U. S. Patent. Schlick and Wurl, No. 944511, 1909.

† In the appendix to an article by Otto Schlick, "The Gyroscopic Effect by Flywheels on Board Ship," Trans. Inst. Naval Architects, 1904, p. 117.

A determination of the constants of a Schlick ship stabilizer by Föppel's method has been given in some detail by Professor John Perry in Nature, **77**, 447 (1908).

Greenhill, Report on Gyroscopic Theory, pp. 34–36.

The Schlick ship stabilizer shortens the period of the ship's roll by increasing the metacentric height by the amount (Art. 86):

$$\frac{h_s^2}{mgK_c}\left[= \frac{2\,K_s}{mK_c}\left(\frac{K_s w_s^2}{2\,g}\right)\right]$$

The quantity within the parentheses represents the rotational energy of the gyro, K_s represents the moment of inertia of the gyro with respect to the spin-axis, and m and K_c refer to the entire ship.

A ship gyro-stabilizer of the inactive type is effective only when the rolling of the ship is periodic. As a matter of fact, the rolling of a ship in a seaway remains periodic but for brief intervals of time. The Schlick device has not been built since the invention of ship stabilizers of the active type (Art. 93).

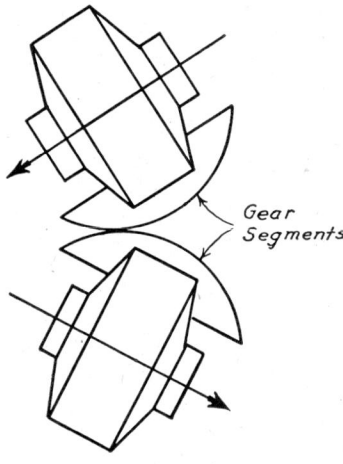

Fig. 117

91. The Fieux Ship Stabilizer. — This consists of two electrically driven gyro-wheels of great moment of inertia spinning in opposite directions about axes that are horizontal and athwartship, Figs. 117, 118. The two gyro-frames are capable of precessing about vertical axes. The two gyro-casings are geared together so that at any instant the velocities of precession of the two gyros are equal and in the opposite direction. This arrangement avoids any effect of the gyros on the steering of the ship. As shown in Art. 89, the stabilizing torques developed by the two gyros are in the same sense about the fore-and-aft axis of the ship.

The outstanding feature of the Fieux* stabilizer is the hydraulic brake mechanism employed to absorb the energy of the ship's roll. This device has the following important properties:

(a) It permits the maximum angle of precession.

(b) It produces zero brake effect when the precession is changing in direction.

(c) It produces maximum opposition to precession when the velocity of precession is maximum and diminishes the brake effect progressively as the precessional velocity decreases.

* Revue Maritime, 1924, p. 351; 1925, p. 180.

THE INACTIVE TYPE OF GYRO SHIP STABILIZER   145

Figure 119 represents a vertical section through the stabilizer in a fore-and-aft plane, and also a horizontal section through ab. The brake mechanism is in a tank forming the base of the appa-

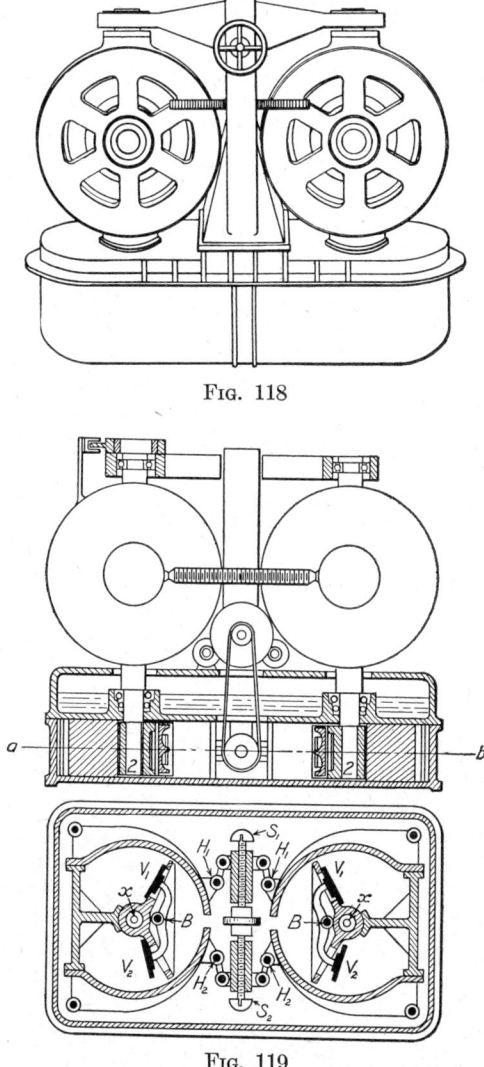

Fig. 118

Fig. 119

ratus. This base is filled with a mixture of glycerine and water. Keyed to the lower end of the vertical precession axle x of each gyro are two vertical vanes V_1, V_2, which move back and forth within a cylindrical enclosure when the gyro precesses. The vanes are pierced by openings controlled by gates connected by an arm pivoted at B. The opposition to the flow of liquid through the openings depends upon the pressure of the gates against the vanes, and this depends upon the speed of precession of the gyro. The braking power is also adjustable by varying the leakage between the vanes and the walls of the enclosing cylinder. Two sections of the enclosing cylinder are hinged at H_1 and H_2. By turning the screws S_1 or S_2, these sections can be moved in or out, thereby changing the opposition offered to the motion of the vanes through the liquid. This adjustment would be made by hand and by an amount depending on the violence of the sea.

§3. *The Active Type of Gyro Ship Stabilizer*

92. The Effect on the Motion of a Swinging Pendulum Produced by an Attached Gyroscope: (b) When the Gyro-Wheel is Acted upon by an Outside Torque about an Axis Perpendicular to the Spin-Axis and the Axis of Vibration of the Pendulum. *Experiment.* — The pendulum shown in Fig. 106 carries a gyro-wheel mounted so as to have two degrees of rotational freedom. The gyro-wheel spins about an axis perpendicular to the axis of vibration of the pendulum and can rotate about a perpendicular axis in the plane of vibration of the pendulum. In the present experiment, the latter axis may be either on, above, or below the center of gravity of the gyro-wheel and attached frame.

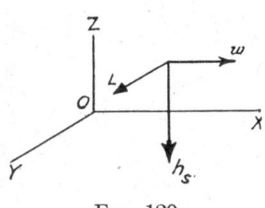

Fig. 120

Hook one end of a stiff wire into an eyelet on the upper part of the frame that carries the gyro-wheel. While the pendulum is at rest, and the gyro-wheel spinning, pull the upper end of the gyro-axle out of the plane of the fixed outer frame. If the gyro-wheel is spinning in the clockwise direction as viewed from above, the pendulum bob will be given a slight motion to the right. This is in accord with the rule given in Art. 36. It is illustrated in Fig. 120, in which the line h_s represents the angular momentum of the gyro-wheel about the spin-axis, w represents the angular velocity produced by the pull on the supporting frame, and L represents the gyroscopic torque thereby produced.

Pull the upper side of the gyro-frame every time the pendulum bob starts to move toward the right from the left end of its path, and push it every time the pendulum starts to move toward the left from the right end of its path. After a few oscillations, the pendulum will be vibrating with a considerable amplitude of swing. The building up of the amplitude of swing is due to the pendulum's being acted upon by a series of separate torques occurring with the same period as the natural period of the pendulum and always in the same direction as the vibration of the pendulum. This is an example of resonance (Art. 26).

Again, while the pendulum is swinging and the gyro-wheel is spinning clockwise as before, pull the upper end of the gyro-shaft every time the pendulum bob starts to move toward the left from the end of its path and push it every time the pendulum bob starts to move toward the right from the other end of its path. This procedure causes the pendulum vibrations to become reduced to zero after a few swings. In these two cases, observe that the direction of the torque applied to the gyro-wheel that increases the amplitude of vibration of the attached pendulum is opposite the direction of the precession of the gyro-axle produced by the motion of the pendulum, and that the direction of the torque applied to the gyro-wheel that damps the amplitude of vibration of the attached pendulum is the same as the direction of the precession of the gyro-axle produced by the motion of the pendulum. Consequently, *in the case of a gyro-wheel having two degrees of rotational freedom mounted on a pendulum in such a manner that the gyro-axle, the axis of precession and the axis about which the pendulum vibrates are mutually perpendicular to one another, the amplitude of vibration of the pendulum will be increased if the gyro-axle be rotated by an outside torque in the direction opposite to the precession, whereas the amplitude of vibration of the pendulum will be decreased if the gyro-axle be rotated by an outside torque in the same direction as the precession.*

In the first case, the periodic gyroscopic torque L, Fig. 120, acting on the pendulum is in phase with the angular velocity of the pendulum. The pendulum is absorbing energy at each vibration and the amplitude of vibration increases (Art. 65). In the second case, the periodic gyroscopic torque applied to the pendulum and the angular velocity of the pendulum are in opposite phase. The pendulum is losing energy at each vibration with a consequent diminution of the amplitude of vibration.

93. The Sperry Ship Stabilizer. — The Sperry ship stabilizer is designed to neutralize each roll increment soon after its inception, thereby preventing the angle of roll becoming large. Its effectiveness is not dependent on the period of roll being constant. It applies the torque developed by the rotation of the spin-axle of a gyro of great moment of inertia, to the damping of the roll of a ship. The gyroscopic torque opposing roll is caused to be in the opposite phase to the velocity of roll. Consequently (Art. 65), the power absorbed from the rolling ship and the damping of roll thereby produced are maxima.

The Sperry ship stabilizer comprises two electrically driven gyroscopes, a powerful electromagnetic brake, an electric motor, two motor generators and a switchboard. One of the gyroscopes

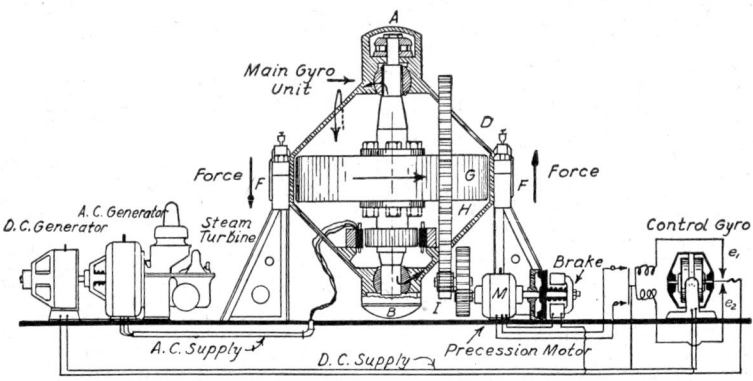

Fig. 121

is small and of three degrees of rotational freedom. The spin-axle precesses when the ship rolls through even such a small angle as a single degree. The other gyroscope is very large and of two degrees of rotational freedom.

The arrangement of the parts of the device as seen on looking from the stern toward the bow of the vessel is as sketched in Fig. 121. The main or stabilizing gyro G, of great moment of inertia with respect to the vertical spin-axis is non-pendulous and is mounted so as to be capable of precession about an athwartship axis. This gyro is spun by alternating current supplied by a turbogenerator as indicated in the diagram. The spin-axle can be rotated forward or back about an athwartship axis by means of the " precession motor " M geared to the casing of the stabilizing gyro. The precession motor is operated by current from a direct current generator.

The direction of rotation of the precession motor and of the connected main gyro-casing is controlled by the small unconstrained gyro. The spin-axis of the control gyro is horizontal and athwartship, in a casing that can rotate about a vertical axis A, Fig. 122, perpendicular to the plane of the diagram. Centralizing springs $b_1 b_2$ are attached to a pin d_1 at one side of the casing. Another pin d_2 attached to the casing can play back and forth between two electric contacts e_1 and e_2. When the pin d_2 is in contact with e_1, a current from the direct current generator rotates the precession motor M in one direction. When contact is made with e_2, the armature of the precession motor rotates in the opposite direction. (In the elevation, Fig. 121, the contacts e_1 and e_2 are shown one above the other. Really, one is behind the other as indicated in the plan view, Fig. 122.)

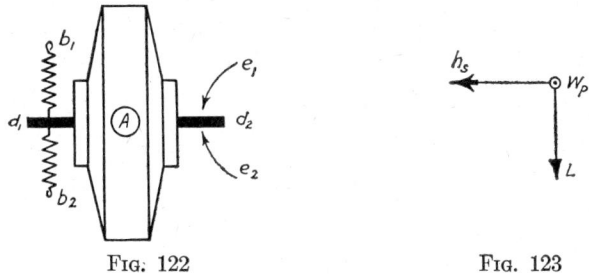

Fig. 122   Fig. 123

The amount and speed of rotation of the spin-axle of the stabilizing gyro G are regulated by a magnetic brake on the armature shaft of the precession motor. This brake has strong springs that seize the armature shaft, thereby preventing rotation of the stabilizing gyro-axle except when the pressure of the springs is opposed by the pull of an electromagnet. The brake coils are in series with the precession motor. Consequently, when current is cut off the precession motor, the magnetic brake seizes the armature shaft and prevents rotation of the stabilizing gyro about the athwartship gudgeons FF, Fig. 121.

94. Operation of the Sperry Ship Stabilizer. — Suppose that the ship rolls to port even so little as one degree of angle, thereby producing on the horizontal spin-axle of the control gyro a torque L, about a horizontal fore-and-aft axis, Fig. 123. Then this axle will precess in the direction w_p, thereby causing the pin d_2 to move over and make contact with e_1. This completes an electric circuit through the precession motor and magnetic brake. As the back electromotive force of the motor when at rest is very small, the initial current through the motor and magnetic brake is large. The brake is completely released, the motor starts very quickly and the stabilizing gyro-axle moves about the athwartship axis with high angular acceleration in the same direction as the precession due to the torque produced by the ship's roll. The magnitude of the angular acceleration depends upon the design of the motor and attached apparatus. About one-half second is required for the control gyro to operate and about one second for the precession motor to acquire full speed.

While the precession motor is gaining in speed, the back electromotive force increases in value, thereby reducing the current in the motor and in the magnetic brake. When the precessional speed of the spin-axle of the stabilizing gyro has the required value, it is maintained practically constant for the required length of time by allowing the proper amount of current to traverse the brake coils.

After the stabilizing gyro-axle has precessed to near the end of its swing, the brake effect is suddenly increased, thereby bringing the precessional speed to zero when the gyro-axle has made the specified maximum displacement from the central position. At this instant nearly one-half period should have elapsed since the spin-axle of the stabilizing gyro was in the central position

When the vessel starts to roll in the opposite direction, the spin-axle of the control gyro precesses in the opposite direction, thereby making an electric contact which causes the precession motor to rotate in the direction opposite to its previous rotation. The same succession of actions above described is now repeated.

The spin-axle of the stabilizing gyro moves back and forth in opposite phase to the velocity of the ship's roll even though the period of roll is not constant. Nearly all the energy that would exhibit itself in the roll of the ship is converted into heat, and the roll is quenched soon after its inception. If any single wave would produce a roll less than that which the stabilizer can prevent, then the control gyro automatically reduces the arc of precession and stops the precession as soon as this single roll has been neutralized. On the other hand, if any single wave would produce a roll greater than that which the stabilizer can neutralize, then the ship will roll through a diminished angle. If the succeeding waves are small, the roll will be entirely quenched

The complete stabilizer equipment of the Sperry active type installed in a large yacht is represented in Fig. 124. It weighs

THE ACTIVE TYPE OF GYRO SHIP STABILIZER 151

about 3 per cent as much as the ship. The Japanese 10,000-ton airplane carrier, *Hosho*, the 2000-ton Italian flotilla leader, *Pigafetta*, and many yachts of various sizes up to 5500 tons are stabilized by similar equipments. The stabilizer for a yacht costs in the neighborhood of one-fortieth as much as the yacht. The largest ship stabilizer ever built is on the Lloyd Sebaudo liner, *Conte di Savoia*. The ship has a displacement of 45,000 tons, metacentric height of 2.2 feet, and period of roll of about 24 sec-

Fig. 124

onds. The stabilizer consists of three complete units each having a main gyro of 215,000 pounds, diameter 13 feet, moment of inertia 4,700,000 pound-feet^2, and speed of 910 revolutions per minute. The entire equipment has a weight of about 624 tons and cost about $700,000. It will prevent the roll of the ship exceeding three degrees from its upright position.

95. The Braking System. — This is a highly important part of any ship stabilizer. The system used in the Sperry active type now will be briefly described. Rolling of the ship produces on the stabilizing gyro a torque that is proportional to the spin velocity of the gyro and the angle of roll of the ship at the given instant.

ANTI-ROLL DEVICES FOR SHIPS

The torques acting on the stabilizing gyro due to rolling of the ship are called *passive moments*. These torques are greatest when the gyro is locked so that it cannot precess.

The gyroscopic torque opposing roll is maximum when the spin-axle of the stabilizing gyro is in the central position. When the spin-axle tilts from that position, the torque diminishes in a manner which depends upon the angle of tilt. Since only small torques are developed when the tilt is large, it is common practice to limit the precession to 60 degrees on each side of the central position. When the stabilizing gyro approaches the end of the permitted angle of precession it causes a " limit switch " to open the direct current generator shunt field. At about the same instant the control gyro also opens the same circuit. Farther precession of the stabilizing gyro drives the precession motor as a generator. Thus, kinetic energy of the stabilizing gyro is first transformed into electric energy and then into heat. This action is called *dynamic braking*. The combination of dynamic braking and the friction braking produced by the magnetic brake can bring the precessional speed to zero in less than one second of time.

The following additional braking effect is possible but is seldom used. When the control gyro releases the magnetic brake, the precession motor armature starts with high acceleration. With increase of precessional speed of the stabilizing gyro there is a decrease in the load on the precession motor. The passive moment due to the rolling of the ship increases the angular speed of the stabilizing gyro about the athwartship gudgeons. The angular speed of the stabilizing gyro about the precession axis may become so great that the precession motor acts as a generator. In case this action occurs, the direction of current in the direct current circuit reverses, thereby causing the direct current generator to operate as a motor. These actions would result in a braking effect that would diminish the velocity of precession. The opposition to the motion of precession produced by the precession motor acting as a generator is called regenerative braking.*

96. Rolling of a Ship Produced by a Gyro. — In certain cases it is desirable to be able to cause a ship to roll. For example, a

---

* Further information regarding the Sperry active type ship stabilizer may be found in the following U. S. Patents:

Sperry, No. 1150311, 1915; No. 1232619, 1917; No. 1452482, 1923; No. 1558514, 1925. Schein, No. 1605289, 1926; No. 1617309, 1927; No. 1655800, 1928.

THE ACTIVE TYPE OF GYRO SHIP STABILIZER

ship aground may be able to free itself if it can roll toward deeper water; a ship stuck in mud may be able to free itself by rolling back and forth till the keel is sufficiently loosened; a ship may be able to break a passage through ice by rolling against the ice sheet.

A vessel equipped with a stabilizer of the active type will roll with gradually increasing amplitude if there be applied to the main gyro a periodic torque that is in opposite phase to the precession of the spin-axle that would be produced by the natural rolling of the ship in the direction in which rolling is desired (Art. 92). By changing electric connections at the switchboard, the apparatus may be caused to operate either as a reducer of rolling or as a producer of rolling.

97. Admiral Taylor's Formula. — The effect of waves on the rolling of a ship is cumulative, each wave contributing a small effect till the total angle of roll may become large. By neutralizing the small increments of roll, the angle of roll will not become large. Admiral D. W. Taylor, U. S. N., has shown that a roll increment ϕ can be neutralized by a gyro-wheel having a moment of inertia relative to its spin-axis of the value

$$K_s' = \frac{1225 \, \phi D H T}{n} \tag{100}$$

where K_s' is expressed in pound-feet^2 units, ϕ is the difference in degrees between two successive amplitudes of roll in the same direction, D is the displacement of the ship in tons, H is the metacentric height in feet, T is the period of roll in seconds, and n is the spin-velocity of the gyro in revolutions per minute. The required moment of inertia may be due either to a single gyro or to two or more gyros.

The derivation of this equation never has been published, but it may be seen in the Archives of the U. S. Navy Department in Washington.

Problem. It is required to compute the principal elements of a ship stabilizer of the non-pendulous active type that will quench a roll increment $\phi = 5°$, where ϕ is the difference between two successive angles of roll in the same direction. The ship has a displacement $D = 2200$ tons, metacentric height $H = 2.5$ ft., period of roll $T = 13$ sec.

The diameter of the gyro-wheel is to be 8 ft. On account of the limit of fiber strength of steel, the peripheral speed must not exceed 33,000 ft. per min. Stops are used that limit the amplitude of precession to 60° on each side of the equilibrium position. A magnetic brake and a motor rotating at 500

154   ANTI-ROLL DEVICES FOR SHIPS

r.p.m. are used to control the precessional speed of the gyro-axle. The full power of the motor together with the torque due to the ship's roll are used to accelerate the precessional velocity for 1.25 sec. after the start of precession. During the first quarter second, the acceleration is 0.3 rad. per sec. per sec. Assume that there is a lag of 0.5 sec. between the inception of a roll and the inception of precession. The precessional velocity attained in about 1.25 sec. after the inception of precession is maintained constant by the use of a motor and brakes for such a length of time that a constant deceleration, then developed, will cause the gyro-axle to come to rest 120° from one end of a swing in the 6.5 sec. of one-half period.

The precession motor is connected to the gyro-casing by gears having diameters in the ratio of 1 to 100. Assume that the efficiency of the gears is 80 per cent, and that the moment of inertia of the gyro-wheel, casing, and gear with respect to the precession axis is nine times that of the gyro-wheel with respect to the spin-axle.

Compute: (a) The moment of inertia and the mass which the gyro-wheel must have if it consists of a uniform disk 8 ft. in diameter; (b) values of the roll velocity of the ship at quarter-second intervals throughout one-half cycle of roll where the maximum amplitude of roll is 2° from the vertical; (c) the horse-power of the motor which, making 500 r.p.m., will produce a mean acceleration of the gyro-axle of 0.3 rad. per sec. per sec. during the first quarter-second of precession; (d) values of the precessional velocity and angular displacement of the gyro-axle at the end of each quarter-second interval throughout one half-cycle; (e) the time at which the final deceleration of the precessional velocity of the gyro-axle should begin; (f) the gyroscopic torque opposing roll at the end of each half-second during one half-cycle of roll.

(g) *Construct* a table with the following data arranged in consecutive columns: t_r, time of roll reckoned from the end of an oscillation; t_p, time of precession reckoned from the end of an oscillation; w_r, instantaneous roll velocity; $\overline{w}_r$, mean roll velocity during the preceding time interval; w_p, instantaneous velocity; ϕ', angular displacement of the gyro-axle from end of oscillation; ϕ_t, displacement of gyro-axle from equilibrium position; $\overline{\Phi}$, mean angular displacement of gyro-axle during preceding time-interval; L_g, gyroscopic torque opposing roll.

(h) *Plot* on the same time-axis, curves coördinating

$$w_r \text{ and } t_r, \quad \phi_t \text{ and } t_r, \quad L_g \text{ and } t_r$$

Solution. (a) The Required Moment of Inertia of the Gyro-Wheel. — From the equation of U. S. Admiral D. W. Taylor, the required moment of inertia of gyro-wheel and axle, with respect to the spin-axis, expressed in pound-feet², is

$$K_s' = \frac{1225 \, \phi D H T}{n}$$

Substituting in this equation the data of the problem:

$$K_s' = \frac{1225(5°)(2200 \text{ tons})(2.5 \text{ ft.})(13 \text{ sec.})}{\frac{33000}{8\pi} \text{ r.p.m.}} = 333350 \text{ lb.-ft.}^2$$

THE ACTIVE TYPE OF GYRO SHIP STABILIZER 155

The mass of a uniform disk 8 ft. in diameter that will have the above moment of inertia, from (22):

$$m \left[= \frac{2 K_s'}{r^2} \right] = \frac{2(333350 \text{ lb.-ft.}^2)}{16 \text{ ft.}^2} = 41670 \text{ lb.} \tag{101}$$

(b) *Velocities of Ship Roll at Instants One Quarter-Second Apart throughout a Roll of Amplitude Two Degrees and Period Thirteen Seconds.* — In the case of a body vibrating with simple harmonic motion of rotation of period T, the angular velocity at time t after leaving the end of an oscillation (41), is

$$w_t = w_e \sin \frac{2 \pi t}{T}$$

where w_e represents the velocity when traversing the equilibrium position. If the amplitude of vibration measured from the equilibrium position be ϕ radians or $\phi°$,

$$w_e \left[= 2 \pi \frac{\phi}{T} = \frac{2 \pi \phi°}{57.3 \, T} \right] = \frac{\phi°}{9.12 \, T} \text{ radians per sec.} \tag{102}$$

Consequently,

$$w_t = \frac{\phi°}{9.12 \, T} \sin \left(360° \frac{t}{T} \right) \text{ radians per sec.} \tag{103}$$

We shall assume that during one roll of a ship, the motion is simple harmonic. Then, if the stabilizer keeps the ship within 2 degrees of the vertical, the velocity of roll at time t_r of the ship considered in this problem is

$$w_r = \frac{2}{(9.12)(13)} \sin \frac{360 \, t_r}{13} = 0.017 \sin (27.7 \, t_r) \text{ radians per sec.} \tag{104}$$

The mean roll velocity during an interval while the instantaneous roll velocity changes at a uniform rate from w_r' to w_r'' is

$$\overline{w}_r = w_r' + \tfrac{1}{2} (w_r'' - w_r') \tag{105}$$

Values of the instantaneous roll velocities at the end of half-second intervals throughout a half-cycle of roll are given in column 3 of the table (p. 161). Values of the mean roll velocities during each half-second interval are given in column 4.

(c) *The Horse-Power of the Motor Which, Making 500 R.P.M., will Produce a Mean Acceleration of the Gyro-Axle of 0.3 Radian per Second per Second During the First Quarter-Second of Precession.* — The mean angular acceleration during any time interval

$$\overline{a} = \frac{\overline{L}_p}{K_p'}$$

where $\overline{L}_p$ represents the mean torque acting upon the precessing system with respect to the axis of the gudgeons about which the system precesses, and K_p' represents the moment of inertia of the same system with respect to the same axis. The total torque is the sum of the torque, L_1, due to the precession motor, the torque, L_2, due to the ship's roll and the torque, L_3, due to the center of mass of the system being not on the axis of precession.

156 ANTI-ROLL DEVICES FOR SHIPS

Since the stabilizer of the present problem is non-pendulous, the torque $L_3 = 0$, and we may write

$$\bar{a}_p = \frac{L_1 + L_2}{K_p'} \tag{106}$$

Since the motor is connected to the gyro-casing by gears of 80 per cent efficiency, and the ratio of the number of teeth on the motor shaft to the number on the gear attached to the gyro-casing is 1 : 100, the torque at the gyro due to the precession motor is

$$L_1 = (0.8)(100) L_1'$$

where L_1' is the torque at the motor.

The power of the precession motor expressed in horse-power is

$$P_h = \frac{2 \pi n L_1'}{33000}$$

whence,

$$L_1' = \frac{33000 \, P_h}{2(3.14)500} = 10.5 \, P_h \text{ at the motor}$$

and

$$L_1 \, [= 80 \, L_1'] = 840 \, P_h \text{ at the gyro}$$

Now we shall find the mean torque L_2 acting on the gyro due to the ship's roll. When the mean roll velocity of the ship is $\bar{w}_r$, and the gyro-axle is in the equilibrium position, the mean torque acting on the gyro has the value,

$$L_2 = K_s w_s \bar{w}_r$$

where K_s is the moment of inertia of the gyro-wheel with respect to the spin-axis expressed in slug-feet^2 and velocities are expressed in radians per sec. If moment of inertia expressed in pound-feet^2 is represented by K_s', and spin-velocity when expressed in revolutions per minute is represented by n, then the preceding equation assumes the form:

$$L_2 = \frac{K_s'}{32.1} \frac{2 \pi n}{60} \bar{w}_r = \frac{K_s' n \bar{w}_r}{307}$$

The mean torque acting on the gyro during the time that the mean velocity of roll is $\bar{w}_r$ and the mean displacement of the gyro-axle from the equilibrium position is $\bar{\Phi}$, has the value

$$L_2 = \frac{K_s' n \bar{w}_r \cos \bar{\Phi}}{307} = \frac{333350(1310)}{307} \bar{w}_r \cos \bar{\Phi} = 1422400 \, \bar{w}_r \cos \bar{\Phi}$$

From column 4 of the table (p. 161), the mean roll velocity $\bar{w}_r$ of the ship while the gyro-axle has precessed for 0.25 sec. from the end of a swing is 0.00504 radian per sec.

The mean angular displacement $\bar{\Phi}$ of the gyro-axle from the equilibrium position during any time interval $(t' - t)$ is

$$\bar{\Phi} = 60 - \tfrac{1}{2} \, (\phi_1' + \phi_2') \tag{107}$$

where the quantities within the parenthesis represent the values of the instantaneous angular displacement of the gyro-axle at the beginning and end of the time interval. At zero time, $\phi_1' = 0$. At time 0.25 sec., ϕ_2' will not be greater than one degree. Suppose that we assume it to be one degree. Whether this assumed value be half the correct value or two times that value

THE ACTIVE TYPE OF GYRO SHIP STABILIZER 157

will make an inappreciable difference in the computed value of the horse-power required to produce the required acceleration of the gyro-axle. Also, as in the subsequent calculations, we shall use a rounded-off number for the horse-power instead of the computed value, it will be safe to assume in the present computation that at time 0.25 sec. $\phi_1' = 1°$. In this case:

$$\overline{\Phi} = 60° - \tfrac{1}{2}(\phi_1' + \phi_2') = 60 - \tfrac{1}{2}(\phi + 1)° = 59.5° \qquad (108)$$

Hence,
$$L_2 = 1422400\, w_r \cos \overline{\Phi} = 1422400(0.00504)0.507 = 3635 \text{ lb.-ft.}$$

From the assumptions of the problem,
$$K_p'[= 9\, K_s'] = 9(333350) \text{ lb.-ft.}^2 = 3{,}000{,}000 \text{ lb.-ft.}^2$$

and the acceleration of the gyro-axle during the first quarter-second is $a = 0.3$ radian per sec.

Substituting in (106), these values of a_p, L_1, L_2 and K_p', we have

$$0.3 = \frac{(840\, P_h + 3635)32.1}{3{,}000{,}000}$$

Whence, the power required of the precession motor is
$$P_h = 29 \text{ H.P.}$$

In the subsequent computations we shall use 30 H.P.

(d) *Values of the Angular Velocity and Displacement of the Gyro-Axle at the End of Each Quarter-Second Interval throughout One Half-Cycle.* — Representing the velocity at the beginning and at the end of a time interval t by w_0 and w_p, respectively,

$$w_p[= w_0 + at] = w_0 + \left(\frac{L_1 + L_2}{K_p'}\right)t = w_0 + \left(\frac{840\, P_h + 1422400\, w_r \cos \overline{\Phi}}{K_p'}\right)t \qquad (109)$$

For the first quarter-second interval $t = 0.25$ sec., $w_0 = 0$ and
$$\overline{w}_r = 0.00504 \text{ rad. per sec.}$$

Hence,
$$w_p = 0 + \left(\frac{25200 + 1422400(0.00504)\cos \overline{\Phi}}{3{,}000{,}000}\right)(0.25)32.1 \text{ rad. per sec.}$$

We shall now find the mean angular displacement $\overline{\Phi}$ of the gyro-axle from the equilibrium position during this interval. Our method will be to make a guess of the angle moved through during this interval and then check the accuracy of the guess. First we shall test the guess that during this time interval the gyro-axle precesses one degree. In this case the mean displacement from the equilibrium, from (108), would be 59.5° and the angular velocity at the end of the interval would be, from the preceding equation

$$w_p = 0.077 \text{ rad. per sec.}$$

An inappreciable error will be made if we assume that during this brief time interval the angular acceleration is constant. In that case, the displacement from the end of a swing would be

$$\phi' = \phi_1' + \tfrac{1}{2}w_p t \text{ radians}$$
$$= 0 + \tfrac{1}{2}[0.077(0.25)57.3]° = 0.55°$$

and the mean angular displacement from the equilibrium position would be
$$\overline{\Phi} = 60 - \tfrac{1}{2}(0 + 0.55)° = 59.72°$$

This value is slightly different from that obtained from the assumption in (108). Our guess of one degree displacement during the first quarter-second was too high. The displacement was more nearly 0.55°. If now we go through a computation as above with 0.55 as the assumed value of ϕ', we shall find that this value will be checked very closely. This is the value to be used.

In the same manner we find values for instantaneous precessional velocity at the end of each quarter-second interval during the first 1.25 sec. of precession, the angular displacement of the gyro-axle from the end of an oscillation and from the equilibrium position at the end of each quarter-second interval, and the mean angular displacement during these quarter-second intervals. The values found are given in columns 5, 6, 7, and 8 of the table.

After 1.25 sec. from the beginning of precession, the brake is applied so as to maintain the precessional velocity constant till the instant at which the velocity is to be given a constant deceleration sufficient to bring the gyro-axle to the end of the 120° swing in 6.5 sec. from the time it was at the other end of the swing. While the velocity of precession is constant, that is, from $t_p = 1.25$ sec. till the instant when the final rapid deceleration is started, values of ϕ', ϕ_t, and $\overline{\Phi}$ can be obtained from the equations:

$$\phi' = \phi_1' + w_p t\, 57.3° \qquad (110)$$
$$\phi_t = 60° - \phi' \text{ degree} \qquad (111)$$
$$\overline{\Phi} = 60° - \tfrac{1}{2}(\phi_1' + \phi_2') \text{ degree} \qquad (112)$$

For example, when $t_p = 1.25$ sec.:

$$\phi' = 14.78° + 0.4333(0.25)57.3 = 20.99°$$
$$\phi_t = 60° - 20.99° = 39.01° = 39° 1'$$
$$\overline{\Phi} = 60 - \tfrac{1}{2}(20.99° + 14.78°) = 42.12° = 42° 20'$$

(e) *The Time at Which the Final Deceleration of the Precessional Velocity of the Gyro-Axle Should Begin.* — As stated in the problem, the angular velocity of the gyro-axle 1.25 sec. after leaving one end of a swing is to be maintained constant until the gyro-axle has come to near the other end of the swing. When the gyro-axle has rotated for the proper time, the brake effect is suddenly increased, thereby bringing the velocity to zero at the end of the 6.5-sec. half-period. We shall now determine the number of seconds after the inception of precession, when a constant deceleration must begin in order that the gyro-axle may attain a displacement of 120° from the other end of the swing in 6.5 sec.

From column 5 of the table (p. 161), we see that 1.25 sec. after the inception of precession, or 1.75 sec. after the inception of the roll, the angular velocity of the gyro-axle is 0.4333 rad. per sec. This is represented by the point B, Fig. 125. Suppose that the constant deceleration must begin at some point C, which is t_1 sec. later than B. Represent by t_2 the time interval from C to the end of the 6.5-sec. half-cycle at D, when the gyro-axle has traversed 120° to this end of a swing. Then, reckoning time from the inception of the roll:

$$t_1 + t_2 = (6.5 - 1.75) \text{ sec.} = 4.75 \text{ sec.}$$
$$t_1 = 4.75 \text{ sec.} - t_2$$

THE ACTIVE TYPE OF GYRO SHIP STABILIZER 159

From column 6 of the table (p. 161), the angle through which the gyro-axle moves during the time $t_1 + t_2$ is

$$120.0° - 14.78° = 105.22° = 1.836 \text{ rad.}$$

Since the deceleration is uniform, the mean velocity of precession during the time t_2 is $\frac{1}{2}$ (0.4333) rad. per sec. Hence,

$$1.836 \text{ rad.} = (0.4333)t_1 + \tfrac{1}{2}(0.4333)t_2$$
$$= (0.4333)(4.75 - t_2) + (0.2166)t_2 = 2.058 - 0.2166\, t_2$$

Consequently,

$$t_2 = \frac{(2.058 - 1.836) \text{ rad.}}{0.2166 \text{ rad. per sec.}} = 1.0 \text{ sec.}$$

Therefore, the final deceleration of the precession of the gyro-axle must be started $(6.5 - 1.0)$ sec. $= 5.5$ sec. after the inception of the roll or 5.0 sec. after the inception of precession.

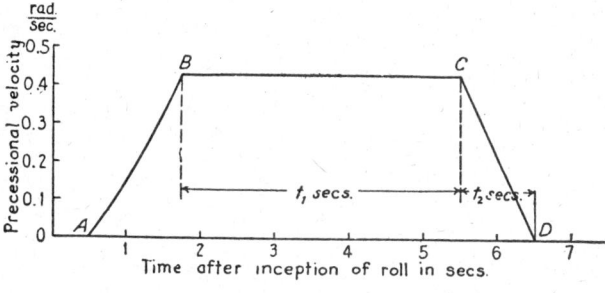

FIG. 125

If t_2 comes out with a negative sign, we would know that with the present constant angular velocity of the gyro-axle during the middle part of a half-cycle, the gyro-axle would need to precess for longer than the half-period before it would reach the end of an oscillation. In this case we would need to increase the velocity during the middle of the half-cycle by allowing the initial acceleration to last for a longer time than that allowed in the problem. Again, if t_2 is positive, but so small that the final deceleration would need to be completed in too short a time, then an impracticable brake torque may be required. In this case, also, we would allow the initial acceleration to last for a fraction of a second longer than that previously allowed.

The value of t_2 should be from 0.5 sec. to 1.0 sec. For a gyro-wheel of the size here used, the value of t_2 we have obtained is larger than strictly necessary. We could have stopped the initial acceleration a little sooner.

(f) *The Values of the Gyroscopic Torque Opposing Roll at the End of Each Half-Second During One Half-Cycle of Roll.* — When the precessing gyro-axle is in the equilibrium position, the torque exercised by the gyroscope on the ship is

$$L_g[= K_s w_s w_p] = \frac{K_s'}{32.1} \frac{2\pi n}{60} w_p = \frac{(333350)1310}{307} w_p = 1422400\, w_p$$

When the precessing gyro-axle is inclined at the angle ϕ_t to the equilibrium position, the gyroscopic torque opposing the ship's roll is

$$L_g = 1422400\, w_p \cos \phi_t \tag{113}$$

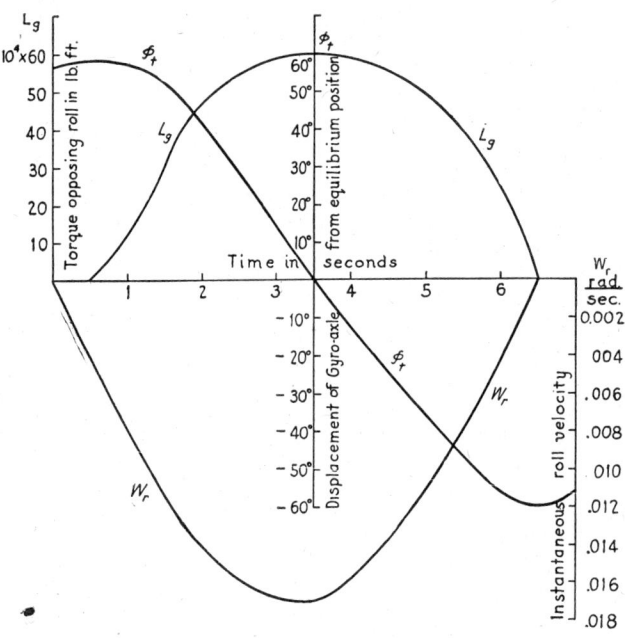

Fig. 126

On substituting in this equation values of the precessional velocity and displacement of the gyro-axle from the equilibrium position as given in columns 5 and 7, we obtain the values of gyroscopic torque opposing the ship's roll given in column 9 of the table (p. 161). These values are plotted against time after the inception of roll, in the curve, Fig. 126.

THE ACTIVE TYPE OF GYRO SHIP STABILIZER

1	2	3	4	5	6	7	8	9
Time from end of roll (starting with ship at maximum angle)	Time of precession measured from max. displacement (Precession lags 0.5 sec.)	Instantaneous roll velocity (Eq. 104)	Mean roll velocity during preceding time interval (Eq. 105)	Instantaneous precessional velocity (Eq. 109)	Angular displacement of gyro-axle from end of oscillation (Eq. 110)	Displacement of gyro-axle from equilibrium position (Eq. 111)	Mean angular displacement of gyro-axle during preceding time interval (Eq. 112)	Gyroscopic torque opposing roll (Eq. 113)
t_r	t_p	w_r	$\overline{w_r}$	w_p	ϕ'	ϕ_t	$\overline{\Phi}$	L_g
seconds	seconds	rad. per sec.	rad. per sec.	rad. per sec.	degrees	degrees	degrees	lb.-ft.
0.00	....	0.00000	0.00	....	....	....	....	....
0.25	....	0.00205	0.00102	....	....	....	....	....
0.50	....	0.00406	0.00301	0.0000	0	60	0	0
0.75	0.25	0.00602	0.00504	0.0770	0.55°	59° 27′	59° 43′	5.6×10.4
1.00	0.50	0.00788	0.00695	0.1581	2.24°	57° 46′	58° 37′	12.00
1.25	0.75	0.00962	0.00875	0.2439	5.12°	54° 53′	56° 19′	20.00
1.50	1.00	0.01129	0.01045	0.3353	9.27°	50° 44′	52° 49′	30.20
1.75	1.25	0.01272	0.01200	0.4333	14.78°	45° 13′	47° 59′	43.4
2.00	1.50	0.0140	0.01336	0.4333	20.99°	39° 1′	42° 20′	47.9
2.50	2.00	0.0159	0.0149	0.4333	33.41°	26° 35′	32° 48′	55.1
3.00	2.50	0.0169	0.0164	0.4333	45.83°	14° 10′	20° 23′	59.7
3.50	3.00	0.0169	0.0169	0.4333	58.25°	1° 45′	7° 58′	61.6
4.00	3.50	0.0159	0.0164	0.4333	70.67°	−10° 40′	−4° 28′	60.5
4.50	4.00	0.0140	0.0149	0.4333	83.09°	−23° 5′	16° 53′	56.6
5.00	4.50	0.01129	0.01264	0.4333	95.51°	−35° 31′	29° 18′	50.2
5.50	5.00	0.00804	....	0.4333 uniform accel.	107.29°	−47° 17′	41° 24′	41.8
6.50	6.00	0.00000	....	0	120°	−60° 00′	....	00.0

CHAPTER V

NAVIGATIONAL COMPASSES

§1. *The Various Types*

98. The Altitude Azimuth Method of Locating the Geographic Meridian. — From the known position of a celestial body, together with simple astronomical observations, the direction of the true north at the place of observation can be determined. Accurately predicted positions of various celestial bodies as well as the times at which certain celestial phenomena will occur are given in Nautical Almanacs published annually by the governments of our maritime nations. Useful tables and all the standard methods of determining navigational quantities are given in Bowditch's American Practical Navigator published by the Hydrographic Office of the U. S. Navy.

The *geographic meridian* of a place on the earth is the great circle of the earth that passes through the given place and the poles of the earth. This line is in the true north-south direction at the given place. The line in the direction of the horizontal component of the magnetic field of the earth at any given place is called the *magnetic meridian* at that place. At very few places is the magnetic meridian in the plane of the geographic meridian. A magnetic needle unaffected by any force except that due to the earth's magnetic field sets itself in the plane of the magnetic meridian at the given place. The *compass bearing* of any object is the angle at the center of the compass card between the magnetic axis of the compass needles and the straight line from the center of the card to the given object.

A celestial body may be located by two quantities. There are three pairs of such quantities commonly employed to locate an object in the sky. They are called altitude and azimuth, declination and right ascension, celestial latitude and celestial longitude.

The celestial sphere is an imaginary sphere of infinite radius onto which, to an observer on the earth, the celestial bodies appear to be projected. Figure 127 represents the celestial sphere, drawn as though it were of finite radius. The small circle represents the earth with the poles marked N and S, respectively. The celestial poles are the points (XX') at which the prolongation of the earth's axis NS intersects the celestial sphere. The celestial equator or equinoctial is the great circle ($QEQ'W$) formed by the intersection of the celestial sphere and the plane of the earth's equator. The great circle H_nEH_sW is the celestial horizon for an observer at A. The celestial meridian, declination circle, or hour circle of any celestial body Y is the great circle of the celestial sphere (XYX') passing through the given body and the celestial poles. The celestial meridian of the place A on the earth is $ZXQ'X'Q$. The point Z at which a vertical line from a place A on the earth intersects the celestial sphere is called the zenith of the given place. The half of a celestial meridian which lies on the same side of the equinoctial as the zenith is called the upper branch; the other half is called the lower branch.

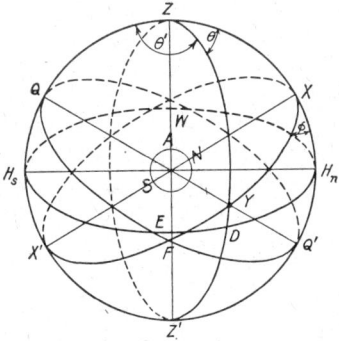

Fig. 127

At a given place A, the polar angle or hour angle of a celestial body Y is the angle ϕ at the celestial poles between the meridian of the place and the meridian of the celestial body. The azimuth or true bearing of a celestial body Y is the angle between the meridian of the observer at A and the vertical great circle passing through the body. In Fig. 127 the azimuth θ' of the body Y measured from the north is $X'ZY$, while measured from the south it is $X'Z'Y$. Azimuths are to be reckoned from the north in north latitudes and from the south in south latitudes. The altitude of any body is its angular distance YD (or angle YAD) from the horizon of the observer, measured upon the vertical great circle through the given body. The zenith distance of a body is its angular distance YZ (or angle YAZ) from the zenith, measured on the vertical great circle through the body. Zenith distance is the complement of the altitude. The declination of a body Y is its angular distance FY (or angle FAY) from the equinoctial, measured on the celestial meridian of the given body. It is designated as north ($+$) or south ($-$), according to the direction of the body from the equinoctial. The polar distance of a celestial body Y is its angular distance YX (or angle YAX) from the pole, measured on the celestial meridian passing through the given body. If the polar distance and the declination are measured from the same pole, the polar distance equals 90° − declination; if they are measured from opposite poles, the polar distance equals 90° + declination.

The hour angle of the sun relative to some chosen place on the earth is called the local apparent time at the given place. Apparent time is expressed either in hours or in degrees. Thus, we may speak of the sun being a certain number of hours, minutes and seconds east (or west) of Greenwich. One hour equals 15 degrees of arc. As the apparent motion of the sun relative to the earth increases and decreases in the course of a year, it is convenient to think of a fictitious "mean sun" that is assumed to have a uniform motion relative to the earth. The Civil Day begins at the instant of transit of the mean sun across the lower branch of the meridian of the observer, that is, at midnight. Clocks indicate civil time, that is, the hour angle of the mean sun, measured from the lower branch of the meridian of some selected place on the earth — Greenwich, for example. At any instant, the difference between the apparent and the mean time, that is, the difference between the hour angles of the apparent and the mean sun, is the equation of time. Equations of time are tabulated in the Nautical Almanac for every even hour of Greenwich civil time throughout the year. Knowing his longitude from Greenwich, an observer with a clock keeping Greenwich civil time can find his local civil time. Then, from the equation of time given in the Nautical Almanac, he can find his local apparent time.

Whenever the sun is visible, the compass bearing can be observed. If we know the three sides of the spherical triangle XYZ, we can compute the azimuth or true bearing θ of the sun, measured from the north, by means of one of the standard equations of spherical trigonometry,

$$\cos^2 \tfrac{1}{2} \theta = \frac{\sin \tfrac{1}{2}(x+y-z) \sin \tfrac{1}{2}(x+y+z)}{\sin x \sin y} \qquad (114)$$

where x, y, and z represent the sides of the triangle opposite the corners X, Y, and Z, respectively. The difference between the computed azimuth and the observed compass bearing of the sun is the angle between the north-south lines of the compass card and the geographical meridian where the compass is situated.

From Fig. 127 we find the following values for the quantities in this equation:

$$\sin \tfrac{1}{2}[x + y - z] = \sin \tfrac{1}{2}[(90° - \text{latitude}) + (90° - \text{altitude}) - \text{polar distance}]$$
$$= \sin [90° - \tfrac{1}{2}(\text{lat.} + \text{alt.} + \text{p.d.})]$$
$$= \cos \tfrac{1}{2} (\text{lat.} + \text{alt.} + \text{p.d.})$$
$$\sin \tfrac{1}{2}[x + y + z] = \sin [90° - \tfrac{1}{2} (\text{lat.} + \text{alt.} + \text{p.d.}) + \text{p.d.}]$$
$$= \cos \left(\frac{\text{lat.} + \text{alt.} + \text{p.d.}}{2} - \text{p.d.}\right)$$
$$\frac{1}{\sin x} = \frac{1}{\sin (90 - \text{lat.})} = \frac{1}{\cos \text{lat.}} = \sec. \text{lat.}$$
$$\frac{1}{\sin y} = \frac{1}{\sin (90 - \text{alt.})} = \frac{1}{\cos \text{alt.}} = \sec. \text{alt.}$$

Substituting these values in (114):

$$\cos^2 \tfrac{1}{2} \theta = \cos \tfrac{1}{2} (\text{lat.} + \text{alt.} + \text{p.d.}) \cos \left(\frac{\text{lat.} + \text{alt.} + \text{p.d.}}{2} - \text{p.d.}\right)$$
$$\times (\sec. \text{lat.}) (\sec. \text{alt.}) \qquad (115)$$

The computation of the left-hand member can be simplified by expressing θ in times of its supplement θ', that is, by reckoning azimuths from the south instead of from the north. Thus,

$$\cos^2 \tfrac{1}{2} \theta = \cos^2 \tfrac{1}{2} (180° - YZX') = \cos^2 (90° - \tfrac{1}{2} \theta') = \sin^2 \tfrac{1}{2} \theta'$$
$$= \tfrac{1}{2} \text{versin } \theta' = \text{hav} (180 - \theta).$$

Half versines of angles, or "haversines" as they are called, are tabulated in Bowditch's American Practical Navigator and in other books on navigation.

99. The Directive Tendency of a Magnetic Compass. — A magnetic compass needle tends to set itself in the direction of the horizontal component of the magnetic field where it is situated. The magnetic poles of the earth do not coincide with the geographic poles. The magnetic north pole is situated in Boothia Peninsula, Canada, at latitude about 70° N., longitude about $96\tfrac{3}{4}°$ W. — more than a thousand miles from the geographic north pole. The magnetic south pole is at latitude about $73\tfrac{1}{2}°$ S., longitude about $147\tfrac{1}{2}°$ E. The number of degrees of angle between the geographic meridian at a particular place and the axis of a compass needle free to turn in a horizontal plane is called the magnetic declination at the particular place.

A line connecting all adjacent points on the earth at which the magnetic declination is zero is called an agonic line. One agonic line is an irregular curve which in the western hemisphere extends from longitude about 96° W., at latitude 70° N., to longitude about 28° W., at latitude 70° S.; and in the eastern hemisphere extends from longitude about 28° E., at latitude 70° N., to longitude about 138° E., at latitude 70° S. A compass needle on an agonic line places itself in the geographic meridian, that is, points true north and south. At various places on the earth, a compass needle that is uninfluenced by any magnetic field except that of the earth will show declinations as great as 180 degrees. In the state of Maine the compass points about 25 degrees west of the geographic meridian, and in the state of Washington it points to the east an equal amount. The compass declination at any point on the earth changes with time. At New York harbor the declination now is about 11 degrees and is increasing at the rate of 6 minutes per year. The magnetic declinations are known for all parts of civilized lands and navigated areas. They are not known, however, for large areas within the arctic and antarctic zones.

The directive tendency of a magnetic compass depends upon the magnitude of the horizontal component of the earth's magnetic field. At no place within either the arctic or the antarctic zone is the horizontal component greater than about one-half the value in New York. Within considerable areas it is nearly zero.

100. The Deviations of a Magnetic Compass on an Iron Ship. — While a ship is being built, the magnetic field of the earth causes the iron structure to become a big magnet with the north-seeking pole toward the north. The various hammering operations facilitate the magnetic induction. If the ship is heading north while being built, there will be developed a north-seeking pole at the lower part of the bow and a south-seeking pole at the upper part of the stern. The steel parts of the ship will retain a part of this magnetism after the ship is launched and is pointing in any direction. A ship that was built with the keel north and south will give no deviations due to this subpermanent magnetism when pointing either north or south. As the ship is rotated 360 degrees, the compass will deviate to the east in one semicircle and westerly in the other (Fig. 128). Deflections of the compass from the magnetic meridian due to this cause are called semicircular deviations. In many cases, semicircular deviations amount to as much as 20 degrees. The semicircular deviation of a compass changes with change of geographical position. If the bow of the ship has been toward the east when building, a north-seeking pole would be developed on the port side. The permanent magnetism of the steel of the ship would give deviations represented by Fig. 129.

The soft iron of a ship becomes magnetized by the earth's magnetic field in which it is situated. As a ship is headed in different directions, the magnetic poles induced in horizontal masses of iron change their position relative to the ship. The compass deviations due to this temporary magnetism are easterly while the head of the ship is pointing in the quadrant from magnetic

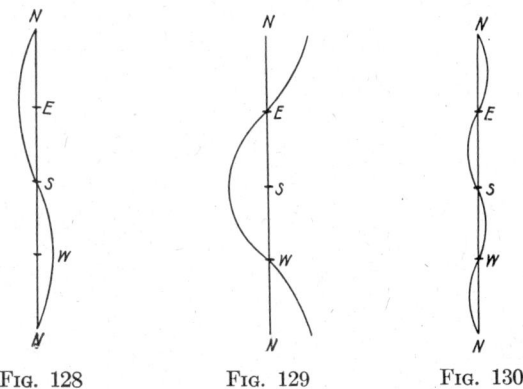

Fig. 128   Fig. 129   Fig. 130

north to east (Fig. 130), and also while the ship is pointing from magnetic south to west. They are westerly while the ship is pointing from east to south and also when it is pointing from west to north. These deviations are called quadrantal deviations. In many cases, quadrantal deviations are as much as 10 degrees. The quadrantal deviation of a compass depends upon the direction of the fore-and-aft axis of the ship relative to the magnetic meridian at the place where the ship is situated.

When an iron ship either rolls or pitches, the compass is affected by changes in the vertical induction of elongated soft iron masses as well as by changes at the compass of the subpermanent magnetic field of the steel parts of the ship. This so-called "heeling error" may amount to more than one degree of compass deviation per degree rolling or pitching of the ship.

A large part of the total deviation of the compass due to fixed masses of iron and steel can be eliminated by properly placed bar magnets, soft iron rods and spheres in the neighborhood of the needle. The residual deviations can be determined experimentally and plotted on a curve for the use of the navigator. However, as the direction and intensity of the earth's field are subject to variations and as the magnetic condition of an iron ship is not constant, the compass indications must be checked frequently against the direction of the geographic meridian as given by the pole star or other celestial body. The directive force acting on the needle of an adjusted magnetic compass is very feeble on the deck of a war ship and is nearly zero within a submarine. It is profoundly altered when large guns and turrets are changed in position.

101. The Deviation of a Magnetic Compass Produced by a Rapid Turn. — If a hardened steel needle be balanced on a pivot, and afterwards magnetized, the north-seeking end will dip below the pivot when the needle is in northern latitudes. To use the magnetized needle as a compass it is kept horizontal by the addition of a counterpoise to the south-seeking end. The center of mass of the needle is no longer vertically below the pivot but is toward the south-seeking end of the needle. If the compass needle is on an airplane making a rapid turn, the center of mass of the needle will lag behind the north-seeking end, thereby causing the north-seeking end to move in the direction the airplane is turning. Thus, when an airplane is executing a rapid turn, the angle of turn indicated by the compass readings is less than the angle actually turned.

If an airplane while moving northward in northern latitudes executes a rapid turn eastward, the compass indicates that the airplane is pointing west of the true course. The angular speed of the needle may be greater than the angular speed of the airplane. In this case, the airplane will appear to be turning westward when really it is turning eastward.

An airplane cannot make a turn unless a force be applied toward the center of the curve. To produce this centripetal force the airplane is tilted about a fore-and-aft axis, the underside of the wings being directed away from the center of the curve. This operation is called " banking." In case the angle of bank, that is, the tilt of the wings from the horizontal, is insufficient to develop the required horizontal thrust against the wings toward the center of the curve, the airplane will not follow a circular path but will slide off along a diverging spiral. When this occurs, the north-seeking end of the compass needle will be deflected in the direction opposite that in which the airplane is turning.

While an airplane is making a turn, the magnetic compass may deflect through a considerable angle either in the direction of the turn or in the opposite direction. The needle may even spin in one direction if the turn is sudden and the angle of banking is correct, or in the opposite direction if the angle of banking is much too small.

102. The Earth Inductor Compass. — A magnetic compass on the instrument board of an airplane is subject to deviations due to proximity to the large mass of magnetized steel and unmagnetized iron of the motor. It is also subject to errors when the airplane makes a turn (Art. 101). Close attention is required to observe the compass card indication — to distinguish between an indication of 43 degrees from one of 48 degrees, for example. It would be better, especially in long-distance flying, if an easily seen index would show simply whether the airplane were on the desired course, or were to the right or to the left.

The earth inductor compass was devised to avoid deviation due to the iron of the motor and also to diminish the difficulty and uncertainty in observing indications of the course. It is essentially a direct-current generator consisting of a coreless armature rotating in the magnetic field of the earth and connected to a millivoltmeter on the instrument board. The armature is placed so far astern that it is outside of the magnetic field due to the iron of the motor. It hangs pendulum-wise, with shaft vertical, from a universal joint connected to the vertical shaft of either an air-driven or an electrically driven motor. The brushes can be turned around the commutator by rotating a dial on the instrument board. The controller dial is graduated in degrees.

When the axis of commutation, that is, the line joining the points of contact of the brushes and the commutator, is perpendicular to the magnetic meridian, no electromotive force is induced in the rotating armature and the indicating millivoltmeter needle is in the zero position. When the line joining the brushes is turned about the armature shaft from this position, the indicating instrument deflects to one side or to the other depending upon the direction in which the brush holder was turned. If it be desired, for example, to fly on a course 30 degrees east of north, the pilot sets the controller disk at 30 degrees east of north and thereafter maintains the plane in the direction that will cause the indicator to remain on the zero mark.

103. The Magneto Compass. — Greater sensitivity than is possible in the earth inductor compass is obtained in Tear's magneto compass,* in which elongated pole pieces, P_1 and P_2, of high

* General Electric Review, xxxii, 1929, p. 190.

permeability and low coercive force, are placed on opposite sides of the armature A, Fig. 131. The armature is rotated by a motor about an axis fixed with respect to the airplane. The longitudinal axis of the pole pieces passes through the center of the armature and is maintained horizontal however the airplane may tilt. The axis of the pole pieces can be turned about a vertical axis by means of a flexible cable connected to a graduated dial on the instrument board.

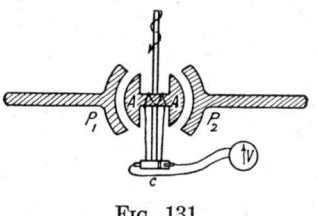

Fig. 131

When the axis of the pole pieces is perpendicular to the magnetic meridian, no electromotive force is induced in the armature and the indicating millivoltmeter needle is at the zero position. When the axis of the pole pieces is not in this position, the needle of the indicating instrument will be deflected. The deflection is independent of the position of the brushes on the commutator C.

With the magneto compass, a course is set and maintained exactly as with the earth inductor compass.

104. The Sun Compass. — In regions where deviations of the earth's magnetic field from the geographic meridian are unknown, and in regions where the intensity of the horizontal component of the earth's magnetic field is too weak to control the direction of a magnetic compass needle, the sun compass is available for indicating directions so long as the sun is visible. The Bumstead sun compass consists of a clock with a single hand that makes the circuit of the face once in twenty-four hours, Fig. 132. The clock is mounted so that it can be turned about horizontal pivots set in two brackets attached rigidly to a horizontal disk. The edge of the disk is marked off into degrees and cardinal points like a compass card. The disk with the attached clock can be turned in azimuth with respect to a lubber line on the base plate. The clock hand carries a shadow-pin and screen by means of which the clock can be turned till the hand points to the sun. At points in the arctic or antarctic zones, the sun remains continuously above the horizon for many days at a time. Suppose that a sun compass in the arctic zone is on the meridian of Greenwich, with the clock face parallel to the sun's rays, the clock set for Greenwich sun time and the hand pointing toward the sun, A, Fig. 133. Then the noon-midnight line of the clock face will be in the meridian plane of Greenwich. While the earth is rotating about its axis in the counter-clockwise direction as viewed by an observer above the north pole, it carries the instrument into the positions A, B, D, F. Throughout the twenty-four hours of a complete rotation of the

Fig. 132

earth, the clock hand continues to point toward the sun so long as the instrument is on the meridian of Greenwich and the clock is set for Greenwich time.

When the instrument is on any meridian and the clock is set for the local time of that meridian, then if the instrument be turned so that the clock hand points toward the sun, the noon-midnight line will lie in the meridian plane of the place where the instrument is situated, C and E, Fig. 133. At any particular place, the clock face must be inclined to the horizontal at an angle of (90° − latitude). To facilitate the making of this setting, there is a divided arc attached to the clock marked off to give this difference for all latitudes at which the instrument would be employed.

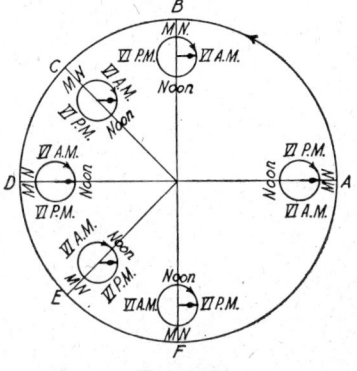

Fig. 133

In using the instrument, the clock is set for local sun time, the face inclined to the horizontal at the proper angle and the clock with the attached disk turned with respect to the lubber line through an angle equal to that of the desired course from the meridian. Then, the proper course is maintained by steering the airplane or ship so that the shadow of the shadow-pin is maintained along the axis of the clock hand.

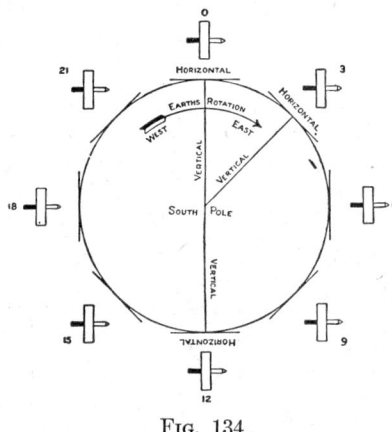

Fig. 134

When the course is not exactly north or south, the apparent time changes. In this case either the clock may be reset after each change of a few degrees longitude, or it may be set permanently to the apparent time of the middle meridian.

105. The Apparent Motion of the Spin-Axle of an Unconstrained Gyroscope Due to the Rotation of the Earth. — According to the First Law of Gyrodynamics (Art. 36), the spin-axle of an unconstrained spinning gyro, unacted upon by any torque, remains fixed in space. This law is also called the law of rigidity of plane of the gyro, or the law of the fixity in space of the spin-axle. If the gyro-axle of a spinning gyro at the equator is parallel to the geographic axis of the earth, then the axle will remain horizontal and in the meridian plane. If the gyro-axle of a spinning gyro at the equator is horizontal in the east-west position, as at O, Fig. 134, then, although the gyro-axle preserves its direction in space as the earth rotates, it appears to an observer on the earth to make one complete turn each twenty-four hours about the horizontal axis in the meridian plane.

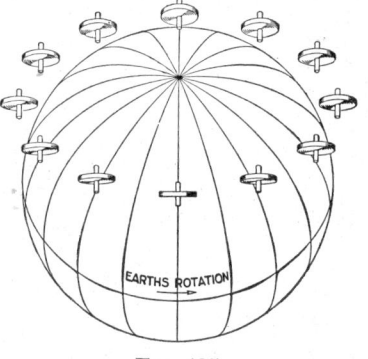

Fig. 135

If an unconstrained spinning gyro be situated between the equator and either pole, then the spin-axle will appear to an observer on the earth to move about both a horizontal and a vertical axle, Fig. 135. At two times during twenty-four hours, the gyro-axle is horizontal and at two times it is in the meridian plane. The motion of the spin-axle relative to the earth is about a cone having the center of the gyro as apex.

The spin-axle of an unconstrained gyroscope continues to be directed toward the same fixed star but it does not continue to be directed toward the same fixed point on the earth. The fixity of the spin-axle in space of an unconstrained gyroscope is inadequate for the production of an instrument that will indicate directions on the earth.

106. The Meridian-Seeking Tendency of a Pendulous Gyroscope. — It has been shown that if a gyroscope be rotated about an axis about which the turning of the gyro-axle is prevented (Art. 49), the axle will set itself parallel to the axis of rotation with the spin of the gyro-wheel in the same direction as the rotation of the gyroscope. It might be imagined that this effect would be adequate for causing the axle of a spinning gyro-wheel, with one degree of angular freedom suppressed, to set itself parallel to the earth's axis. For example, if a spinning gyro-wheel on the earth is mounted so that angular motion about every horizontal axis is suppressed, the spin-axle will tend to set itself parallel to the earth's axis and will turn toward the meridian plane. If, however, the instrument is on a moving ship, the spin-axle will tend to set itself parallel to the axis of the resultant of the angular velocity of the earth and that of the ship with respect to the earth. Since the angular velocity of the ship is often much greater than that of the earth, a gyro-wheel mounted in this manner would not give correct indications. To be of value for use as a compass, the gyro-axle (a) should be urged toward the meridian by a torque sufficient to bring it into the meridian within a reasonable length of time after being set into spinning motion, (b) should quickly return to the meridian when displaced therefrom, (c) should be nearly horizontal when in the meridian.

If, at some instant, the spin-axle of a gyroscope north of the equator is nearly horizontal and in the meridian plane, then at succeeding instants the north-seeking end will have an apparent motion away from the meridian toward the east. In order that the gyro-axle may maintain its position in the meridian, the north-seeking end must be given a westerly precessional velocity equal to the vertical component of the earth's angular velocity. This required precession can be produced by a torque about a horizontal axis which tilts the gyro-axle from the horizontal plane. Two different methods are now in use to produce this tilting torque on the spin-axles of gyro-compasses. One is by making the sensitive element pendulous. In the following Article the other method will be considered in which the tilting torque is produced by a moving mass of liquid.

Now it will be shown how the weight of a mass attached to the supporting frame of a spinning gyro-wheel below the center of the gyro-wheel, will cause the spin-axle to precess toward the meridian in which the gyroscope is situated. Consider a gyroscope with horizontal axle pointing east and west, X, Fig. 136. In this diagram, it is imagined that we are looking down on the northern hemisphere of the earth as from an airship. Suppose that the freedom of the gyroscope about the horizontal axis is restricted by hanging a mass m on the lower side of the supporting frame. We shall call a mass suspended from the supporting frame of a gyro-wheel "a pendulous mass." The pendulous mass will be pulled toward the center of the earth. This pull will produce zero torque when the gyro-axle is horizontal as at X. While the earth rotates, the spin-axle of the gyro-wheel tends to maintain its position in space with the result that, when the gyroscope has reached a position Y, the axle is no longer horizontal. The weight F of the pendulous mass exerts a torque in the counter-clockwise direction about an axis perpendicular to the plane of the diagram. The line L, representing the torque, is directed upward from the plane of the diagram. Suppose that the direction of spin is as represented by the line h_s. Then, from the law of precession, the spin-axle tends to become parallel to the torque-axis with the direction of spin in the direction of the torque. In the present case, the direction of the torque is the same as the direction of rotation of the earth. Therefore, the gyro-axle of the pendulous gyroscope tends to become parallel to the earth's axis, with the

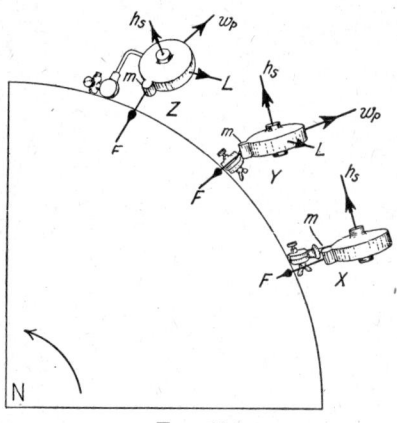

Fig. 136

direction of spin in the same direction as the rotation of the earth. In the same manner it can be shown that if the center of mass of the gyroscope is above the point of support, the gyro-axle tends to become parallel to the earth's axis, with the spin in the opposite direction to the rotation of the earth.

When the spin-axle reaches the meridian plane (position Z, Fig. 136), the north-seeking end of the gyro-axle is at its maximum elevation above the horizon and the weight of the pendulous mass is exerting its maximum torque thereby producing a maximum velocity of precession. The gyro-axle will cross the meridian

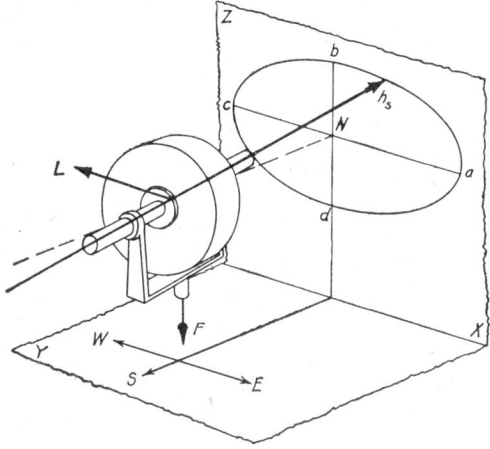

Fig. 137

plane and, continuing its angular motion beyond the meridian plane, the axle will dip and pass through the horizontal position. During the dipping, the torque due to the weight of the pendulous mass decreases till it is zero when the spin-axle is horizontal. At this instant, the precessional velocity is zero and the spin-axle is at its greatest angular displacement from its original direction at X. Continuing its dipping, the north-seeking end of the spin-axle passes through the horizontal plane, raising the pendulous mass and thereby developing a torque in the opposite direction. The north-seeking end of the spin-axle rises, again crosses the meridian plane and repeats its motion back and forth. The oscillation of the spin-axle of a gyro-compass back and forth through the meridian plane involves an exchange of energy between the compass and the earth.

It has now been shown that the deflection of the spin-axle of a pendulous gyroscope from the horizontal develops a torque that produces a precession of the spin-axle toward the meridian plane. If the spin-axle be displaced out of the meridian plane, the north-seeking end will oscillate back and forth around an elliptical orbit *abcda*, Fig. 137. This figure represents the path of the prolongation of the gyro-axle on a vertical plane as seen by an observer looking from the south toward the north. If the motion be undamped, the angular amplitude of each oscillation will equal the original angular displacement from the meridian. The period of vibration depends upon the torque acting on the oscillating system, the angular speed of the spinning gyro-wheel and upon the moment of inertia of the oscillating system. It is always made to be about 84 minutes (Art. 113).

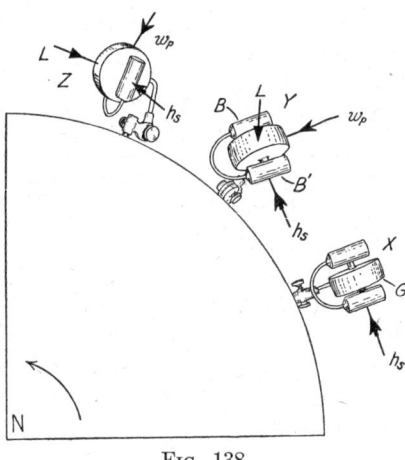

Fig. 138

107. The Meridian-Seeking Tendency of a Liquid-Controlled Non-Pendulous Gyroscope. — Attached to the sides of the frame supporting the gyro-wheel, G, Fig. 138, are reservoirs joined by a small tube.* This figure represents the view as seen by an observer above the north pole of the earth. With the gyro-axle horizontal, the two reservoirs are filled with mercury, up to the level of the middle of the gyro-axle. The center of gravity of the gyroscope is raised to the center of gravity of the gyro-wheel by means of an adjustable counterpoise above each reservoir. This mercury-filled system is called a "mercury ballistic."

Suppose that at a particular instant the spin-axle is horizontal and east and west as shown at X, Fig. 138. In Fig. 137, a line along the spin-axle intersects the vertical plane XZ at the point a, east of north. While the gyroscope is carried by the rotation of the earth to the position Y, Fig. 138, the spin-axle tends to preserve its direction in space, thereby causing the reservoir B to be raised above B' relative to the horizontal plane at B. Consequently, some mercury flows from the reservoir B to the reservoir

* U. S. Patent. Harrison and Rawlings, No. 1362940, 1920.

B', thereby producing a torque about a horizontal axis perpendicular to the plane of the figure and directed away from the reader. This torque produces a precession that causes the spin-axle to tend to set itself parallel to the torque-axis and with the direction of spin in the direction of the torque. As this torque continues, the spin-axle becomes parallel to the meridian plane, the tilt of the spin-axle becomes maximum and the direction of spin is opposite the direction of rotation of the earth. A line along the spin-axle now intersects the vertical plane at b, Fig. 137. The spin-axle crosses the meridian plane and becomes less tilted from the horizontal. The spin-axle becomes horizontal when the inclination to the meridian plane is maximum. A line along the spin-axle now intersects the vertical plane at c. Owing to the rotation of the earth, the dip will continue past the horizontal and mercury will flow toward the reservoir at B, thereby causing a precession in the reverse direction. During this precession, a line along the spin-axle traces a curve cda on the vertical plane.

The cycle of motions causes the end B' to trace an elliptical path with major axis horizontal, east and west, and minor axis vertical. The end B is the north-seeking end of the gyro-axle. If there be no damping, this motion is repeated with undiminishing amplitude. The period is made to be about 84 minutes (Art. 113). From the preceding consideration it is seen that the direction of spin of the liquid-controlled gyroscope is opposite to the direction of rotation of the earth, whereas the direction of spin of the pendulous gyroscope is the same as the direction of rotation of the earth.

108. Making a Gyroscope into a Gyro-Compass. — In Arts. 106 and 107 it has been shown that, owing to the rotation of the earth, the spin-axle of a gyroscope, with rotational freedom about a horizontal axis partially suppressed, will trace the surface of an elliptical cone. The apex of the cone is at the center of mass of the gyro. The axis of the cone is horizontal and in the meridian plane. The maximum amplitude of deviation of the spin-axle from the meridian may be many degrees. If the gyroscope were on the earth or on a stationary ship, the meridian could be located by taking the mean of the extreme positions of the spin-axle. The length of time required for this determination is so great that the gyroscope, as described, is quite useless for determining the course of a ship at sea.

To make a gyroscope into a gyro-compass, means must be provided to damp the oscillations to such an extent that very quickly the spin-axle will move into and remain in the meridian plane. The required damping can be effected by a torque that will either (a) diminish the tilt of the spin-axle from the horizontal, or (b) oppose the horizontal motion of the spin-axle. The oscillation of the spin-axle of a gyro spinning in the same direction as the rotation of the earth can be damped by means of a viscous liquid. The magnitude of the damping torque should be proportional to the instantaneous angular speed of the gyro-axle about the axis of the torque.

109. The Meridian-Seeking Torque Acting on a Gyro-Compass.
— The spin-axle of a gyro-compass is kept approximately horizontal by either a pendulous mass or a mercury ballistic. Freedom of rotation of the gyro about a horizontal axis perpendicular to the spin-axle is partially suppressed by a torque that is developed when the spin-axle is turned. If a gyroscope with one degree of rotational freedom either wholly or partially suppressed be rotated, the spin-axle will tend to set itself in the direction of the axis of rotation (Art. 49). Consequently, the spin-axle of a stationary gyro-compass on the earth tends to set itself parallel to the horizontal component of the earth's angular velocity at the place where the compass is situated.

When the spin-axle and the axis about which the instrument is being rotated are perpendicular to one another, the torque tending to turn the gyro-axle has the value (Art. 36):

$$L = K_s w_s w_e = h_s w_e \tag{116}$$

where w_e represents the angular velocity of the earth about the polar-axis.

The magnitude of the torque urging the spin-axle toward the meridian plane when the gyro-compass is at latitude λ, and the spin-axle is inclined to the meridian plane at the angle ϕ, is obtained by substituting for w_e the value of the component of the angular velocity of the earth with respect to a horizontal line perpendicular to the vertical plane that contains the spin-axis. In Fig. 139, the spin-axle of the gyro is in the direction AO, with respect to a system of rectangular coördinates consisting of a vertical VO, a horizontal line MO in the meridian, and a horizontal east-and-west line EO. A vertical plane containing the spin-axle intersects the horizontal plane MOE in the line HO. The vertical plane VOH is inclined to the meridian plane VOM at the angle ϕ.

At a point O on the earth at latitude λ, Fig. 140, the component angular velocity of the earth with respect to a horizontal axis OB in the meridian plane at O is $w_e \cos \lambda$. In Fig. 139, this component is represented by the line OB. The component of OB about a horizontal axis PB perpendicular to the vertical plane VOH containing the spin-axle of the gyro-wheel is

$$OB \sin \phi = w_e \cos \lambda \sin \phi$$

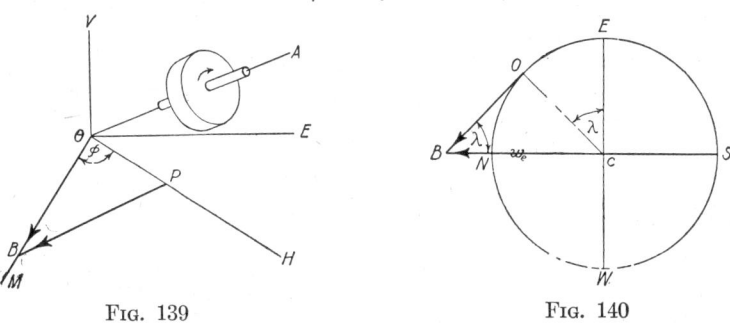

Fig. 139   Fig. 140

Substituting this value for w_e in (116) we have

$$L = h_s w_e \cos \lambda \sin \phi \tag{117}$$

This is the magnitude of the torque urging the spin-axle of a stationary gyro-compass toward the meridian. It depends upon the angular momentum of the gyro, upon the latitude and upon the angle between the spin-axle and the meridian. The meridian-seeking torque is so small, especially when the gyro-compass is at a place of high latitude and the spin-axle makes a small angle with the meridian, that unusual methods must be adopted to reduce the opposing torque due to the method of suspending the sensitive system.

For the passage under the ice to the North Pole, Sir Hubert Wilkins' *Nautilus* was equipped with a gyro-compass. Now the directive tendency of the sensitive element of a gyro-compass produced by a constant precession toward the meridian is too small to be effective at latitudes greater than about 85 degrees. Consequently, when near the Pole, the compass gyro was disconnected from the precessing device and the gyro left as free of all restraint as possible. In this condition, the spin-axle holds its direction in space instead of maintaining its position in a meridian of the earth. The spin-axle moves away from the meridian, to which the spin-axle was parallel at the instant the gyro was disconnected from the precessing device, with an angular velocity given by (68), Art. 42.

The spin-axle of a gyro-compass is preferably horizontal, because (a) the directive force is proportional to the horizontal component of the angular momentum of the gyro, (b) the compass is used to measure bearings, that is, angles on a horizontal plane. If the spin-axle were inclined to the horizontal, it might be possible to project the direction of the axle on a horizontal plane — if we could assume a given plane to be horizontal. The uncertainty in the horizontality of a given plane would introduce an appreciable error in the determination of either bearings or the meridian plane.

A gyro-compass having a gyro of large angular momentum is less affected by small disturbing torques than is one having a gyro of smaller angular momentum.

If the compass be on a vehicle that could move with such a linear velocity that its angular velocity about the earth's axis is equal and opposite the angular velocity of the earth, then the meridian-seeking torque would be zero. The relation between the linear velocity of the vehicle in knots and the latitude at which the directive tendency of the gyro-axle will be zero can be easily obtained. Thus, remembering that one knot is a speed of one nautical mile per hour and that a nautical mile is $\frac{1}{60}$ of $\frac{1}{360}$ of the earth's equatorial circumference, it follows that a point on the equator moves with a linear speed of $\frac{(360 \times 60)}{24} = 900$ knots. The motion is eastward. At latitude λ, a point on the earth has a velocity of 900 cos λ knots eastward. So that a gyro-compass will have zero directive tendency when on a vehicle moving westward at v knots at latitude λ if

$$v = 900 \cos \lambda$$

For example, if an airship is flying westward at 90 knots, the gyro-compass will have zero directive tendency at latitude λ given by the equation $90 = 900 \cos \lambda$. In this case $\lambda = 80° 10'$.

It will now be shown that, on account of the tilt of the spin-axle when in the resting position, any great change in the spin-velocity of the gyro will result in a deflection in azimuth of the spin-axle. The gyro-wheel constitutes the rotor and the gyro-casing constitutes the field magnet of an electric motor. When the gyro is spinning, there is a reacting torque applied to the casing. So long as the spin-velocity is constant this torque is balanced by

THE VARIOUS TYPES 181

windage and frictional torque; but if the spin-velocity should change, the torque acting on the casing is unbalanced.

If the spin is in the same direction as the rotation of the earth (pendulous gyro-compass), then when the spin-velocity is accelerating there will be a torque acting on the gyro-casing tending to turn the casing in the direction opposite that of the rotation of the earth. If the north-seeking end of the gyro-axle is tilting upward, the vertical component of this torque will produce a slight easterly deviation of the north-seeking end of the spin-axle. When the spin-velocity of the same compass is decelerating, there will be a westerly deviation.

In the case of a gyro-compass having the spin in the direction opposite to the rotation of the earth (mercury ballistic compass), changes in the spin-velocity result in deviations in the directions opposite those for the pendulous gyro-compass.

PROBLEMS

1. The Brown gyro-compass has a single gyro-wheel of 4.5 lb. wt., diameter 4 in., spinning at 15,000 r.p.m. Assuming that the radius of gyration is 0.74 of the radius, find the meridian-seeking torque, in grain-inches, at latitude 40° when the spin-axle is inclined 1° to the meridian.

2. The Sperry Mark VI gyro-compass has a single gyro-wheel of 54 lb. wt., diameter 10 in., spinning at 6000 r.p.m. Assuming that the radius of gyration is 0.75 of the radius, find the meridian-seeking torque in grain-inches at latitude 40° when the spin-axle is inclined 1° to the meridian.

3. The Sperry Mark X compass has a single wheel of moment of inertia 22.2 lb.-ft.^{2}, spinning at 10,000 r.p.m. Find the meridian-seeking torque in grain-inches at latitude 40° when the spin-axle is inclined to the meridian at 1°.

4. One model of Arma compass has two gyro-wheels rotating in the same direction with axles inclined to one another at 60°, the apex of the angle being toward the south. Suppose that each gyro-wheel weighs 5 lb. 2 oz., has diameter of 5.12 in. and spins at 20,000 r.p.m., and that the radius of gyration is 0.74 of the radius of the wheel. Find the meridian-seeking torque in grain-inches when the compass is at latitude 40° and bisector of the angle between the two spin-axles at an angle of 1° to the meridian.

5. An early model of Anschütz gyro-compass consisted of three gyro-wheels each of mass 5 lb. 2 oz., diameter 5.12 in., spinning at 20,000 r.p.m. These gyro-wheels were arranged at the apexes of an equilateral triangle. The axle of the south gyro-wheel made equal angles to the sides of the triangle while the axles of the other two gyros were along the sides of the triangle. A determination gives 0.13 lb.-ft.^{2} for the moment of inertia of each gyro about its spin-axle. Find the meridian-seeking torque of the system in grain-inches when at latitude 40° and the axis of the south gyro is inclined 1° to the meridian.

§2. *The Natural Errors to which the Gyro-Compass is Subject*

110. The Latitude Error. — Owing to the rotation of the earth, combined with the tendency of the spin-axle of a gyroscope to maintain its direction in space, the end of the axle of a gyro-compass moves back and forth across the meridian plane and also up and down across the horizontal plane through the center of the gyro-wheel (Arts. 106, 107). If the gyro-compass is north of the equator and if the gyro-axle is horizontal and pointing east of the meridian plane, the north-seeking end of the spin-axle will rise and move westward. The tilting of the spin-axle is necessary for the production of the torque required to cause the gyro-axle to seek the meridian. At any latitude the tilt of the spin-axle of an undamped compass, when the axle is in the meridian plane, is just sufficient to produce the rate of precession about a vertical axis required to maintain the axle in the meridian.

Any damping action that opposes the natural tilting of the spin-axle will result in a lower rate of precession than that required to maintain the axle in the meridian. The resting position of the spin-axle will then be deviated from the meridian plane by an angle called the *latitude error* or the *damping error*. If the method used for damping does not involve an opposition to tilting of the gyro-axle, then there will be zero latitude error. When there is latitude error, its magnitude depends upon the latitude and upon the constants of the compass. The Florentia and the Sperry gyro-compasses are subject to a latitude error. This latitude error is compensated automatically by a device that forms part of the compass.

111. The Error Due to the Velocity of the Ship. The Meridian-Steaming Error. — A gyro-compass on board a ship at sea is subject to deflecting forces due to the motion of translation of the ship and also to angular motions of steering, rolling and pitching. The deflection of the spin-axle from the meridian due to the linear velocity of the ship will now be considered.

Suppose that the ship is moving with a linear velocity v_s on a course inclined at an angle θ to the meridian. Then the component in the direction of the meridian is $v_s \cos \theta$. The angular velocity of the ship with respect to the earth about an east-and-west axis has the value

$$_e w_s = \frac{v_s \cos \theta}{R}$$

where R represents the radius of the earth.

NATURAL ERRORS

If we represent the angular velocity of the earth about its axis, with respect to some fixed line in space, by ${}_pw_e$, then the component angular velocity of the earth relative to a horizontal axis in the meridian plane at a place of latitude λ (Fig. 141) is

$${}_mw_e = {}_pw_e \cos \lambda$$

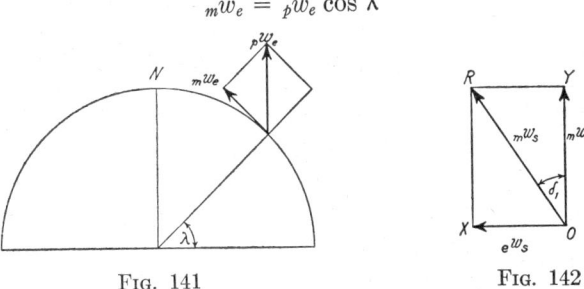

Fig. 141 Fig. 142

In Fig. 142, the two angular velocities ${}_ew_s$ and ${}_mw_e$ are represented by the lines OX and OY, respectively. The resultant angular velocity of the ship with respect to a horizontal axis in the meridian plane, ${}_mw_s$, is represented by OR. The spin-axle of the gyro-compass sets itself parallel to the axis of the resultant angular velocity of the ship. In this figure

$$\tan \delta_1 = \frac{{}_ew_s}{{}_mw_e} = \frac{v_s \cos \theta}{R\, {}_pw_e \cos \lambda} \tag{118}$$

This deflection, δ_1, called the meridian or north-steaming error, is westerly for a northern course, and easterly for a southern course. It is independent of the construction of the compass. In Art. 113 it will be shown how the appropriate meridian-steaming deflection can be produced, without oscillation, just as soon as any change occurs in either the speed or the course of the ship.

If the linear speed of the ship be expressed as v knots, then

$$v_s = 6080\, v \text{ ft. per hour.} \quad R = 20{,}925{,}000 \text{ ft.}$$

$${}_pw_e = 1 \text{ rev. per day} = \frac{2\pi}{24} \text{ radian per hour} = 0.26 \text{ rad. per hour.}$$

Hence,

$$\tan \delta_1 = \frac{6080 \text{ feet per hour}}{(20{,}925{,}000 \text{ ft.})(0.26 \text{ rad. per hr.})} \frac{v \cos \theta}{\cos \lambda}$$

Neglecting the small difference between δ_1 and $\tan \delta_1$, and remembering that 1 radian = 57.3 degrees, we have the value of the meridian-steaming error:

$$\delta_1 = 0.0011 \frac{v \cos \theta}{\cos \lambda} \text{ radians} = 0.063 \frac{v \cos \theta}{\cos \lambda} \text{ degrees} \tag{119}$$

Problem. — A ship at latitude 60° N. is steaming north-east with a speed of 20 knots. Find the deflection of the spin-axle of the gyro-compass from the geographical meridian due to the meridian-steaming error.

Solution. — From (119), since $\theta = 45°$ and $\lambda = 60°$,

$$\delta_1 = 0.063 \frac{20\,(0.707)}{0.5} = 1°.8 \text{ west of the meridian}$$

112. The Deflection of the Axle of a Gyro-Compass Produced by Acceleration of the Ship's Velocity. The Ballistic Deflection Error.

Experiment. — The pivots supporting the inner frame of the gyroscope, Fig. 143, are horizontal. A mass m attached rigidly to this frame and hanging below the gyro-wheel tends to keep the gyro-axle horizontal when the frame is at rest. Place the gyroscope on the intersection of two lines drawn at right angles to one another on the table, and marked S-N and W-E, respectively. Place the spin-axle parallel to the S-N line, and set the gyro-wheel spinning in the direction indicated by the arrow h_s.

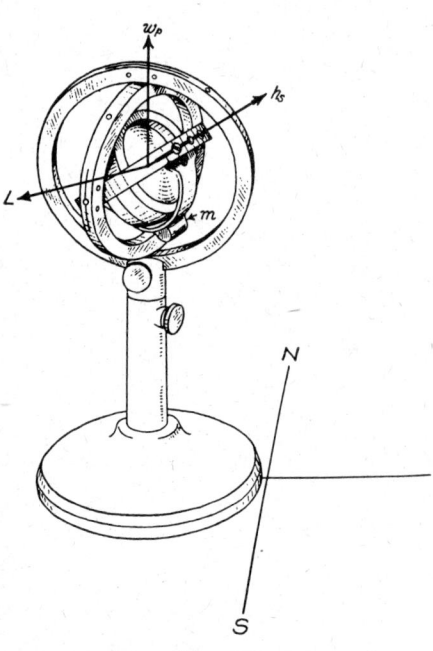

Fig. 143

(a) Slide the instrument quickly along the table in the direction S-N. During the time the motion of the frame is being accelerated in the direction S-N, the pendulous mass m hangs back, thereby developing a torque about an axis parallel to the E-W line in the direction indicated by the arrow L. The spin-axis turns toward the torque-axis about a vertical axis in the direction represented by the arrow w_p.

(b) Accelerate the motion of the gyro-frame in the direction N-S or decelerate the motion in the direction S-N. Note that during the change in velocity the spin-axle precesses in the direction opposite that developed in the preceding case (a).

(c) Push the gyroscope in the direction S-N with constant velocity and then suddenly move it toward either the right or left. Note that, during the change in the direction of the velocity, the spin-axle precesses in the same direction as in case (b) when the motion in the direction S-N was decelerated.

(d) Now accelerate the motion of the gyro-frame in either the direction E-W or W-E, that is, perpendicular to the spin-axis. No change in the direction of the spin-axis is produced.

Question. Suppose that we have a non-pendulous mercury ballistic gyro-compass with the spin of the gyro in the direction opposite that considered above. Find the direction of the deflection of the spin-axle that would be produced if the instrument were to be accelerated in each of the ways considered above.

So long as the velocity of a ship is constant, the spin-axle of the gyro-compass will have a resting position at a small angle from the meridian of a magnitude depending upon the velocity and latitude of the ship (Arts. 110 and 111). If the velocity of the ship is changing while the ship is on any heading except perpendicular to the spin-axle of the gyro-compass, then during the time the velocity is changing, the spin-axle will be deflected from the resting position which the spin-axle had when the ship was moving with its first velocity. This deflection will be constant while the acceleration remains constant. The deflection of the spin-axle from its resting position, produced by an acceleration of the meridian component of the ship's velocity, is called the *ballistic deflection*.

A diminution of velocity in the meridian is produced not only by diminishing the speed in the meridian but also by changing the heading either eastward or westward. In every case, the ballistic deflection produced by any acceleration of the ship's velocity is in the same direction as the change in the direction of the resting position of the gyro-axle, on account of the changing meridian-steaming error.

Consider a ship steaming north at 10 knots. Because of the meridian-steaming error, the resting position of the spin-axle will be at an angle *NOX*, west of the meridian, Fig. 144. Suppose that now the speed on the same heading be increased to 20 knots. The resting position of the spin-axle corresponding to the new velocity is shifted to *YO*. Suppose that during the time the speed was changing from 10 knots to 20 knots, the ballistic torque acting on the gyro-frame caused the gyro-axle to precess into the position

ZO. The ballistic deflection is *XOZ*. The angle *YOZ* by which the gyro-axle either overshoots or undershoots the correct settling position for the particular latitude, course and speed, is called the *ballistic deflection error*. With any change in the meridian component of the ship's velocity, the spin-axle will be deflected in the same direction as the movement of the gyro-axle produced by the change in the meridian-steaming error. When the acceleration of the ship's velocity ceases, the gyro-axle oscillates about the resting position *YO*. There will be zero ballistic deflection error if *XOZ* = *XOY*, that is, if the ballistic deflection equals the change in the meridian-steaming error.

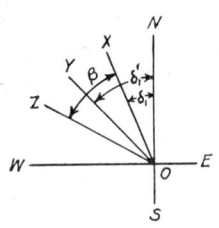

Fig. 144

In order that the compass card may be in its true resting position throughout the time the velocity of the ship is changing, the compass must be forcibly precessed, during this time, to the resting position proper to the speed, course and latitude at the end of the acceleration.

If the gyro were not spinning, the spin-axle would tilt throughout the time the velocity of the ship is accelerating. If, however, the gyro were spinning, the "rigidity of plane" would oppose tilting. In this case, the pendulous mass would be acted upon by a force which would develop precession about a vertical axis. If the precessional velocity were of the proper magnitude, it would cause the compass to precess, during the time the velocity of the ship is accelerating, from the resting position proper to the speed, course and latitude at the beginning of acceleration, to the resting position proper to the speed, course and latitude at the end of the acceleration. Under this condition, there would be zero ballistic deflection error.

113. The Period of a Gyro-Compass That Will Have Zero Ballistic Deflection Error When at a Definite Latitude. — The magnitude of the ballistic deflection depends upon the torque that opposes tilting of the sensitive element. In the case of a gyro-compass of the pendulous type at a given latitude, the torque that opposes tilting depends upon the degree of pendulousness of the sensitive element. In the case of a liquid-controlled non-pendulous gyro-compass at a given latitude, it depends upon the weight of liquid that is moved from one side of the gyro-case to the other and upon the distance between the opposite mercury reservoirs. Each of these factors affects the period of vibration of the spin-axle back and forth through the meridian. It will now be proved that if the period of the sensitive element has a certain value, which for all latitudes is not far from 84 minutes, then the spin-axis of the gyro-compass will move, without oscillation, to the resting position appropriate to the given speed and course, just as soon as any change occurs in either the speed or course.

Imagine a symmetrical body supported at the center of mass so as to be capable of turning about this point in any direction. An acceleration of the motion of the body in any direction will produce no turning of the body.

Imagine, farther, that a mass m be attached to the body at a distance x from the center of mass C, Fig. 145. When the motion of the system is constant, the mass m is acted upon by a force mg vertically downward and by an equal force vertically upward. If the linear motion of the system is subject to a constant acceleration a in a horizontal direction to the right, the pendulous part m of the system must be acted upon by a force ma in the direction of the acceleration. This force is the resultant of the weight mg vertically downward and a force F directed toward the point C. This force F is inclined to the vertical at an angle ϕ, given by the relation

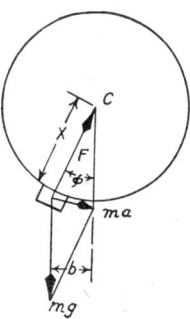

Fig. 145

$$\tan \phi = \frac{ma}{mg} = \frac{a}{g} \qquad (120)$$

When the horizontal distance of m from the vertical line through C is b, the system is acted upon by a torque about an axis perpendicular to the plane of the diagram of the magnitude

$$L = mgb = mgx \sin \phi$$

When ϕ is so small that $\sin \phi$ may be replaced by $\tan \phi$, we have, on substituting the value of $\sin \phi$ from the above equation, in (120):

$$L = mxa$$

Suppose that the suspended system includes a gyro spinning about a nearly horizontal axis. We are now dealing with a pendulous gyro-compass. If the ship carrying the gyro-compass be accelerated while on a meridian course, the pendulous part of the sensitive element will be acted upon by a torque as above. If, however, the course be perpendicular to the spin-axle, the torque will be zero. Representing the meridian component of the mean acceleration of the ship during the short time t by a, there will act upon the gyro a torque L which produces a precessional velocity about the vertical axis of the magnitude,

$$w_p \left[= \frac{L}{h_s} \right] = \frac{mxa}{h_s}$$

In this short time t the meridian component of the velocity of the ship and of the compass changes by an amount at. During this time, the gyro-axle turns through the angle

$$\beta [= w_p t] = \frac{mxat}{h_s} = \frac{mx}{h_s}(v_t - v_0) \qquad (121)$$

Where β is the "ballistic deflection" and $(v_t - v_0)$ is the change in the meridian component of the velocity of the ship in time t.

When the meridian component of the velocity of the ship changes from v_0 to v_t, the resting position of the gyro-axle is deflected through an angle $(\delta_1' - \delta_1)$, from the position δ_1, Fig. 144. Now β depends only upon the constants of the compass and the linear acceleration of the ship, whereas the meridian-steaming errors depend upon latitude and velocity. The ballistic deflection β and the deflections δ_1 and δ_1' due to changes in the meridian-steaming error are in the same direction. Consequently, a gyro-compass can be designed with such a period T_0 that, when the instrument is at some particular latitude, the ballistic deflection shall equal the change in the meridian-steaming error $(\delta_1' - \delta_1)$ produced by the change in the velocity of the ship.

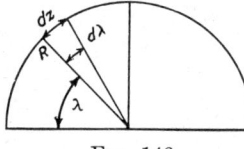

Fig. 146

If the ship, starting from rest, moves along a meridian through a distance dz in time dt, with constant linear acceleration a, the velocity at the end of time t will be

$$\frac{dz}{dt} = at$$

During the motion, the latitude of the ship has changed by the amount, Fig. 146:

$$d\lambda = \frac{dz}{R}$$

where R is the radius of the earth. The rate of change of latitude is

$$\frac{d\lambda}{dt} = \frac{dz}{R\,dt} = \frac{v'}{R} = \frac{at}{R} \qquad (122)$$

where $\frac{dz}{dt}[= at]$ is the meridian velocity v' of the ship at the end of time t. Substituting in (121) the value of at from (122), we obtain for the ballistic deflection

$$\beta = \frac{mxR}{h_s}\frac{d\lambda}{dt} \qquad (123)$$

Before equating this value of the ballistic deflection and the value of the meridian-steaming error, we shall express the latter in terms of the time rate of change of latitude. Since the meridian-steaming error δ_1 is so small that we may neglect the difference between δ_1 and $\tan \delta_1$, (118) may be written

$$\delta_1 = \frac{v'}{Rw_e \cos \lambda}$$

where the meridian component $v_s \cos \theta$ of the velocity of the ship is represented by v' and the symbol ${}_p w_e$ is abbreviated to w_e.

On substituting in this equation the value of v' from (122), we have for the meridian-steaming error

$$\delta_1 = \frac{d\lambda}{w_e \cos \lambda \, dt} \qquad (124)$$

Now equating the ballistic deflection (123) and the meridian-steaming error (124),

$$\frac{MxR}{h_s} = \frac{1}{w_e \cos \lambda}$$

or
$$h_s = mxRw_e \cos \lambda \qquad (125)$$

When the angular momentum has the value given by this equation, the spin-axle will move without oscillation to the resting position proper to the particular latitude and final velocity. On substituting this value in (88) we find that the period of the azimuthal vibration back and forth through the meridian of the undamped compass that produces zero ballistic deflection error has the value

$$T_0 = 2\pi\sqrt{\frac{R}{g}} \qquad (126)$$

At the equator $R = 20,925,000$ ft., and $g = 32.086$ ft. per sec. per sec. Hence, at the equator, $T_0 [= 5068$ sec.$] = 84.4$ min. Since at other latitudes the radius of curvature of the earth is slightly less and g is slightly greater than at the equator, the period of the compass at places either north or south of the equator should be somewhat less than at the equator.

Since (125) must hold when the period of azimuthal vibration of the undamped compass is such that there is zero ballistic deflection error, we see that, at any latitude λ, the period can be adjusted to the desired value by varying either h_s or mx. Hence, a gyro-compass at latitude λ will have zero ballistic error when either h_s or mx is adjusted so that the ratio

$$\frac{h_s}{mx} = Rw_e \cos \lambda \qquad (127)$$

The mx in this equation is called the "pendulous factor."

The period of every gyro-compass is made of such a value that when the instrument is at a selected latitude, the ballistic error is zero. The desired period is obtained by having proper values of the angular momentum of the gyro and of the pendulousness of the sensitive element or the mass of active mercury in the ballistic.

114. The Ballistic Damping Error. — While the meridian component of a ship's velocity is being accelerated, either by a change of speed or of course, there will be a tilting of the gyro-compass except when the acceleration is perpendicular to the spin-axle. This is due to forces impressed by the damping mechanism and to the precession in azimuth produced by the pendulousness. When the degree of pendulousness is correct for a particular latitude, then the gyro-compass will give zero ballistic deflection error at that latitude during the time the ship's velocity is accelerating. At any other latitude, the gyro-compass will show a ballistic deflection error during the time the velocity of the ship is accelerating. The tilt of the spin-axle associated with the acceleration of the ship's velocity is superposed on the tilt due to the rotation of the earth. The torque producing this additional tilt ceases when the acceleration ceases but the spin-axle starts oscillating back and forth through the equilibrium position of the spin-axle for the particular velocity and latitude of the ship.

The oscillation of the gyro-compass may continue for an hour or more after the acceleration of the ship's velocity has ceased. The device employed to damp the oscillation exerts a torque on the spin-axle which displaces the settling point from the normal equilibrium position. The maximum deflection of the spin-axle from the equilibrium position due to this cause is called the maximum *ballistic damping error*. In the case of any gyro-compass, the damping of the vibration of the gyro-axle produces an error after the ship has completed a change in course or speed. This ballistic damping error is most marked after the ship, steaming at full speed, has completed a turn of 90 degrees or more. It attains a maximum value in about 20 minutes after the velocity of the ship has become constant. The magnitude of the damping error increases with increase in the acceleration of the ship. Much greater accelerations are produced by sudden turns than by any possible change in the speed of the ship. Since this error is small while the ship is on a straight course but may be large when the ship is making a turn, it is also called the *ballistic turning error*. It is also called the *damping acceleration error*.

The ballistic damping error and the accompanying oscillation of the spin-axle can be prevented by stopping the operation of the damping device during the turning of the ship.

115. The Compass Error Due to Rolling of a Ship When on an Intercardinal Course. The Quadrantal or Rolling Error. — A pendulous gyro-compass on a ship that is rolling or pitching is acted upon by a force that has a maximum value at the end of a roll or pitch, and another force that has a maximum value when the pendulous system is passing through its equilibrium position. The first of these forces will be considered in the present Article. The second will be considered in the following Article.

A ship's compass is supported in a Cardan gimbal mounting consisting of two horizontal rings, one capable of rotation about an axis parallel to the keel of the ship and the other about a transverse axis. In so far as freedom to turn in any direction is concerned, a gyro-compass on board ship is equivalent to a gyro-wheel mounted in five rings. For simplicity of representation, in the present Article, the pendulous gyro-compass will be represented by a gyro-wheel in a casing free to turn about any axis through its center and carrying an additional mass attached to the lower side of the casing. This mass that causes the gyro and casing to be pendulous we shall call " the pendulous mass."

Consider the effect on the direction of a gyro-compass produced by rolling or pitching of a ship about an axis that is perpendicular to the spin-axle. In Fig. 147, the keel of the ship is east and west, perpendicular to the plane of the diagram — the bow of the ship pointing away from the reader. Suppose that the ship oscillates back and forth about a horizontal axis perpendicular to the plane of the diagram through O. At the end of a swing, the inertia of the pendulous mass m carries the mass beyond the equilibrium position, thereby developing a torque on the gyro-wheel about an axis perpendicular to the plane of the diagram indicated by the sign (∗) or (·) marked on the diagram. If the direction of the spin velocity is that represented by the arrow marked h_s, then, when the compass has reached the end of a swing

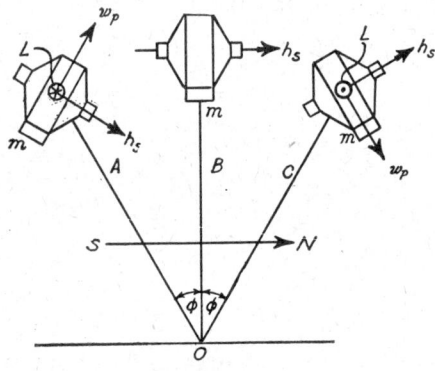

Fig. 147

at A, the north-seeking end of the spin-axle will be deflected away from the reader toward the west. When the gyro-compass comes to the other end of an oscillation at C, the north-seeking end of the spin-axle will be deflected toward the east. A succession of back and forth swings of the ship about a horizontal axis produces a succession of vibrations of the gyro-axle but no permanent deflection. Thus, when the ship is pointing east or west, rolling produces a vibration of the gyro-axle but no fixed deflection from the meridian. Similarly, when the ship is pointing north or south, pitching produces a vibration but no fixed deflection of the gyro-compass. The period of vibration of the spin-axle is that of the oscillation of the ship. As this is never greater than about 15 seconds whereas the period of the entire compass is several minutes, there is no resonant vibration of the spin-axle. Consequently, the amplitude of vibration of the spin-axle due to rolling and pitching is small.

Suppose that the pendulous gyro-compass is on board a ship that is either rolling or pitching about an axis through O parallel to the spin-axle and perpendicular to the plane of Fig. 148. While the ship oscillates back and forth through the equilibrium position, the pendulous mass m moves back and forth with the same period. When the pendulous mass comes to the end of its swing, its inertia carries it beyond its equilibrium position thereby rotating the gyro-casing in the direction indicated by the curved arrow drawn from m. Since this rotation is about an axis parallel to the spin-axle it produces

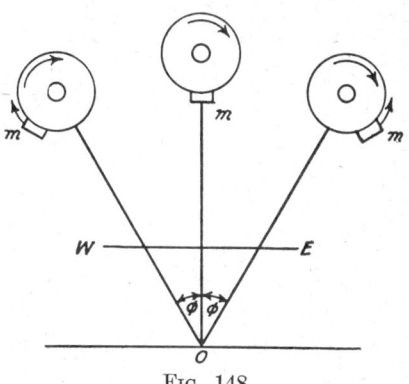

Fig. 148

zero effect on the direction of the spin-axle. Thus, when a ship is on either a north or a south course, rolling produces no deflection of the gyro-compass. When on an east or west course, pitching produces no deflection.

In the preceding paragraphs, it has been shown that when a ship is steaming on a cardinal course, north, east, south, or west, neither rolling nor pitching produces a deviation of the gyro-compass from the meridian. Now, we shall consider the effect on the direction of the spin-axle of a pendulous gyro-compass produced by rolling of the ship when steaming on an intercardinal course. To fix the ideas, suppose that the course is NW-SE as indicated in Fig. 149. The rolling of the ship causes the compass to move back and forth along a path AB, perpendicular to the course, and causes the pendulous mass to move back and forth relative to the true vertical through the center of the gyro. After a roll to the north-east, the center of mass of the pendulous mass will be at a point c east of the true

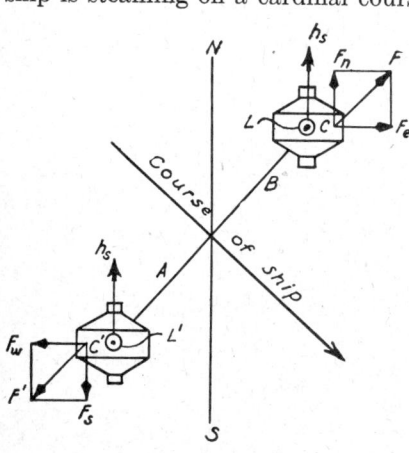

Fig. 149

vertical through the center of the gyro and back of the plane of the diagram. A force F will act at this point in the direction of the roll. After a roll to the south-west, the center of mass of the pendulous mass will be at a point c' west of the true vertical through the center of the gyro and back of the plane of the diagram. A force F' will act as this point in the direction of the roll.

Each of the two forces F and F' may be replaced by three components, one in the north-south line, one in the east-west line and another in the vertical line. The components F_e and F_w, in the east-west line, are equal, oppositely directed, and perpendicular to the gyro spin-axle. These forces produce no net torque on the spin-axle. The vertical components produce torques about the spin-axle but these torques produce no deflection of a horizontal spin-axle.

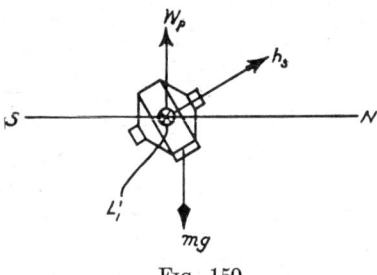

Fig. 150

After a roll to the north-east, the meridian component F_n of the F acting at c has a lever arm equal to the distance between its line of action and the spin-axle of the gyro. It produces a torque L on the gyro, about a true vertical axis through the center of the gyro, in the counter-clockwise direction as seen when viewed from above. After a roll to the south-west, the component F_s of the force F' applied at c' acts on the gyro with a torque L', in the counter-clockwise direction, about a true vertical axis through the center of the gyro. Thus, when the ship rolls in either direction, the gyro is acted upon by a counter-clockwise torque about a true vertical axis.

Since the direction of spin of a pendulous gyro-compass is clockwise when viewed from the south toward the north, and since the spin-axle tends to set itself parallel to the torque-axis and with the direction of spin in the direction of the torque, Art. 106, it follows that the north-seeking end of the spin-axle of the compass here considered will tilt upward while the ship rolls. The upward tilt raises the center of mass of the pendulous mass and moves it to the north of a vertical line through the center of the gyro. Figure 150 represents a view as seen when looking from the east toward the west. The weight mg of the pendulous mass develops a torque L'_1 which, in turn, produces a precessional motion w_p. As a result, the north-seeking end of the spin-axle precesses westward.

It has now been shown that a pendulous gyro-compass on a rolling ship headed either north-west or south-east is deflected toward the west. In the same manner it can be shown that a pendulous compass on a ship headed either north-east or south-west will be deflected toward the east when the ship rolls. The spin-axle of a pendulous gyro-compass on a rolling ship steaming on any intercardinal course is deflected in the direction to make it more nearly parallel to the keel of the ship. The deflection due to pitching is in the direction opposite that due to rolling for a heading in any direction.

116. Quadrantal Deflection Due to Lack of Symmetry of the Sensitive Element. — It has been shown that, if the mass of a swinging pendulum bob is not distributed symmetrically with respect to the pendulum axis, the bob will tend to set itself in the position in which its moment of inertia with respect to the vibration axis is maximum (Art. 14).

In the case of all gyro-compasses having a sensitive element consisting of a single gyro, inert masses must be added to the sensitive element in order that the moment of inertia of the suspended system with respect to every horizontal axis through the center of mass may be equal. The mass, however, of a sensitive element comprising two or more gyros can be distributed symmetrically about the line through the center of mass and the center of oscillation. Neither a compass consisting of a single gyro with properly adjusted balancing masses nor a compass consisting of two or more properly placed gyros is subject to deflections due to centripetal forces developed during rolling and pitching of the ship.

117. The Suppression of the Quadrantal Error. — No feature of the gyro-compass has so taxed the ingenuity of physicists and designers as the avoidance or elimination of the quadrantal error. In Art. 115 it has been shown that the quadrantal error is due to the cumulative effect of vertical components of the precession of the spin-axle produced by the swinging of the compass system from side to side as the ship rolls or pitches. The quadrantal error would be avoided: (*a*) if the suspended compass system remained vertical however the ship rolls or pitches; (*b*) if the impulses imparted to the swinging gyro-frame by the rolling and pitching of the ship were applied only when the line through the center of the gyro-wheel and the center of gravity of the pendulous mass is vertical; (c) if the point of application of the force on the gyro due to the pendulous mass were shifted back and forth from one side of the vertical to the other in such a manner that a vertical precession is thereby produced which is equal and opposite to that due to the rolling or pitching of the ship; (d) if the natural period of oscillation of the pendulous system were much greater than the period of oscillation of the ship; (e) if the sensitive element were non-pendulous.

118. The Degree of Precision of Gyro-Compasses. — In order that a compass may furnish a satisfactory base line for gun-fire control, it must have a greater degree of precision than is required of a compass used only for navigating the ship. Large naval vessels are equipped with gyro-compasses of the most careful design and workmanship known to the art. Such compasses are called " gun-fire control " compasses. The less severe requirements of merchant vessels and naval vessels of the smaller sizes are met adequately by instruments of a somewhat simpler type called " navigational compasses." Of the gyro-compasses mentioned in the following Articles, the Arma Mark IV and the Sperry Mark X instruments are " gun-fire control " compasses. The others are " navigational " compasses.

The tests of gyro-compasses submitted for use in the United States Navy include measurements of the undamped period and the damping percentage of the sensitive element, the change of settling point when the voltage impressed on the gyro-motors changes, and the change of settling point when the binnacle is rotated about a vertical axis or when rocked about various horizontal axes. The tests of the changes of the settling point produced by rotation and rocking of the binnacle are made by means of a machine called a " scoresby," from the name of the inventor. This machine consists of a platform that can be rotated in either direction about a vertical axis and that can be rocked with a given period and with a given amplitude about a horizontal axis in any azimuth. Figure 151 represents a scoresby rocking a gyro-compass G as in an actual test.

If x and y represent two successive amplitudes of swing in the same direction, then $\dfrac{x-y}{x}$ may be called the fractional damping and $100\left(\dfrac{x-y}{x}\right)$ would be called the damping percentage. The damping percentage is calculated from the amplitudes of the first three peaks as follows — divide one hundred times the amplitude of the second peak by the amplitude of the first peak and subtract the quotient from one hundred; divide one hundred times the amplitude of the third peak by the amplitude of the second peak and subtract the quotient from one hundred. The mean of these two results is the mean percentage of damping. If the compass be precessed either toward the east or toward the west 30 degrees from the settling point and then released, the damping percentage must not be less than 60 per cent nor more than 80 per cent.

When rotated about a vertical axis at a uniform rate of 3 degrees per second, a gyro-compass of the "navigational" grade must not show an error greater than 0.5 degree during a run of twenty hours, and a gyro-compass of the "gun-fire control" grade must not show an error greater than 0.2 degree.

When rocked back and forth about a horizontal axis through an angle of 15 degrees from the vertical

Fig. 151

with a period of 9 seconds a gyro-compass of the "navigational" grade must not show an error greater than 0.6 degree during a run of forty-eight hours, whatever may be the heading, and a gyro-compass of the "gun-fire control" grade must not show an error greater than 0.3 degree.

When rocked back and forth about a horizontal axis through an angle of 40 degrees from the vertical with a period of 9 seconds a gyro-compass of the "navigational" grade must not show an error greater than 1.0 degree during a run of forty-eight hours, whatever may be the heading, and a gyro-compass of the "gun-fire control" grade must not show an error greater than 0.5 degree.

When the voltage impressed on the gyro-motor (or motors) is changed from normal to 10 per cent more or less than normal, the maximum settling point error of a gyro-compass of the "navigational" grade must not exceed 0.5 degree, and for a gyro-compass of the "gun-fire control" grade the error must not exceed 0.2 degree.

A gyro-compass equipment includes the master compass, an electric motor-generator, storage battery and switchboard, together with repeater compasses. Many installations include a course recorder, a radio direction finder, and an automatic pilot operated by the master compass. By means of a gyro-compass controlled automatic pilot, the ship may be kept in a straight course as long as may be desired without the aid of a helmsman. The master compass and electric plant are installed between decks. Then the ship is steered from a regular repeater compass, and bearings are taken from two other repeater compasses at the ends of the officers' bridge.

The magnetic compass is a simple and relatively cheap instrument, but the errors to which it is subject can be but partially eliminated or allowed for. On the other hand, the gyro-compass is a complicated and expensive outfit, but the errors to which it is subject can be completely eliminated automatically. A standard magnetic compass for a large ship costs about $500. The cost of a master gyro-compass such as is found on merchant ships and the smaller naval vessel costs about $3000. The cost of a gyro-compass equipment as used on merchant ships, consisting of the master compass, electric plant, three repeaters and course recorder, costs about $5500. An automatic pilot adds about $2200 to the cost of the compass equipment. A single gyro-compass equipment of the highest grade, such as is used for gun-fire control on large battle ships, costs from three to four times as much as one of the " navigational " grade used on merchant ships. It is usual to have two complete " gun-fire control " gyro-compass equipments on a large battle ship.

§3. *The Sperry Gyro-Compass*

119. The Principal Parts of the Master Compass. — Since 1920 all gyro-compasses made by the Sperry Gyroscope Company[*] have been single-wheel instruments of the liquid-controlled non-pendulous type (Art. 107). The various models differ in directive torque, degree of precision, and in details of design, but they are similar in all essential respects. The models commonly employed on mercantile vessels are designated Mark VI and Mark VIII. The gyro-wheel of each is 10 inches in diameter, weighs 54 pounds,

[*] Made in Brooklyn, N. Y., in London, England, and in Tokio, Japan.

THE SPERRY GYRO-COMPASS 199

and rotates in the open air at a speed of 6000 revolutions per minute. The two models differ in that the gyro-wheel of Mark VI constitutes the armature of a direct current motor and the field coils are imbedded in the gyro-casing, whereas the gyro-wheel of Mark VIII constitutes the rotor of a three-phase alternating current motor and the stator coils are imbedded in the gyro-casing. Mark V and Mark X are designed for use on naval vessels. The gyro-wheel of the Mark V is 12 inches in diameter, weighs 45 pounds and rotates at 8600 revolutions per minute. The gyro-wheel of the Mark X Gun-Fire Control Gyro-Compass* is $13\frac{5}{8}$ inches in

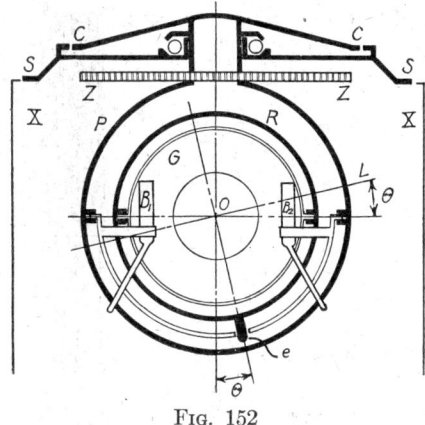

FIG. 152

diameter, weighs 122 pounds and rotates at 10,000 revolutions per minute. The gyro-wheels of Mark V and Mark X rotate in a partial vacuum of about 28 inches of mercury below atmospheric pressure.

A simplified diagram showing the fundamental parts of the Sperry gyro-compass, as seen when looking toward the compass from the south, is represented in Fig. 152. The gyro-wheel, spinning about a nearly horizontal axis perpendicular to the plane of the diagram, is enclosed by the gyro-casing G. The gyro-casing is supported on horizontal bearings by a vertical ring R which, in turn, is suspended by a system of vertical wires from the top of a tube extending upward from an outer ring P called a "phantom." The phantom is supported by ball bearings on a "spider" S supported on gimbal rings carried by the binnacle XX. The phantom carries the compass card C, and the "azi-

* See Frontispiece.

Fig. 153

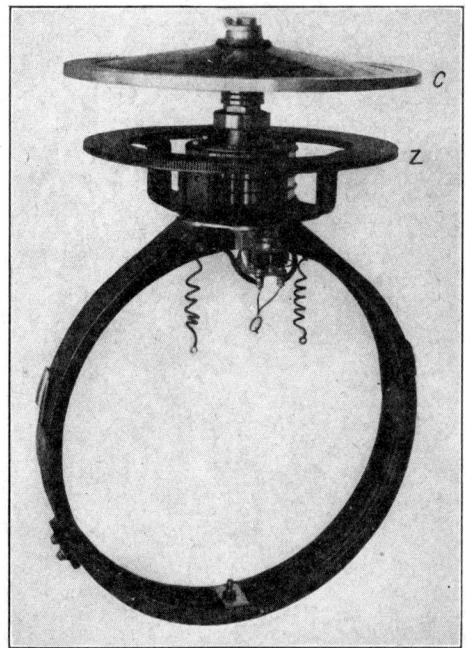

Fig. 154

THE SPERRY GYRO-COMPASS 201

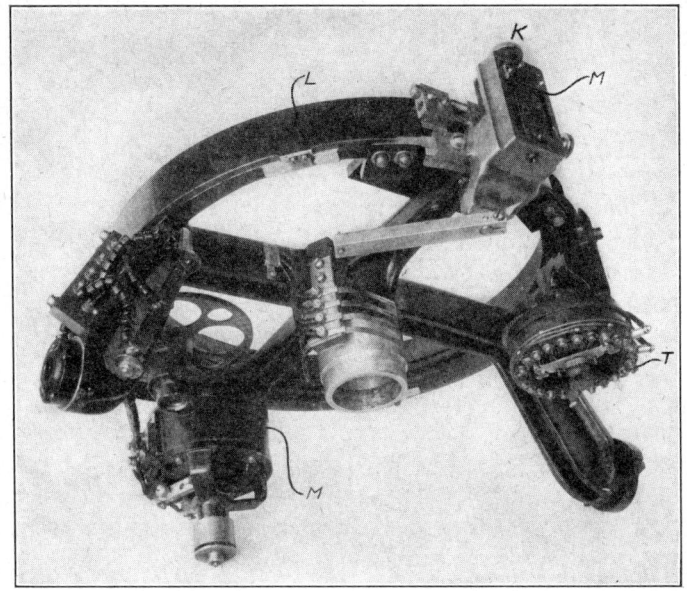

Fig. 155

Fig. 156

muth gear " Z, by means of which the phantom may be rotated so as to bring it into the plane of the vertical ring. The " mercury ballistic " B_1B_2 is fastened to a frame pivotted to this phantom. It is loosely connected to the gyro-casing by a pin e placed a little to the east of the vertical line through the center of mass of the suspended system.

The gyro-wheel, casing and vertical ring constitute the " sensitive element," Fig. 153, which is acted upon by a meridian-seeking torque due to the rotation of the earth. The phantom element is shown in Fig. 154. The supporting spider is shown in Fig. 155. The complete gyro-compass, Mark VIII, removed from the binnacle is shown in Fig. 156. In this compass, the mercury ballistic consists of two pairs of mercury reservoir B_1B_1' and B_2B_2', each pair connected by a small tube.

120. The Follow-Up System. — When the sensitive element becomes turned out of the meridian plane in either direction through an angle even so small as one-quarter of a degree, an electric contact is made which causes a small reversing motor, M, Fig. 155, to rotate the azimuth gear, Z, Fig. 154, through an equal angle in the same direction. Thus the phantom element follows the sensitive element, thereby preventing any appreciable twisting of the suspending wires. In this manner, the suspending wires never develop an appreciable torque in opposition to the meridian-seeking torque acting on the compass. The follow-up system neutralizes the effect of frictional forces that oppose motion about the vertical axis.

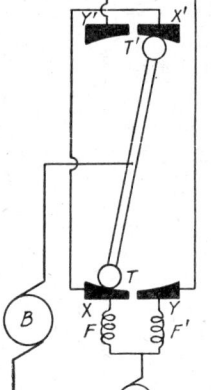

Fig. 157

A pair of contactor plates, insulated from each other, is carried on each of the opposite sides of the phantom element, Q, Fig. 154. Lightly pressing against each pair of contractor plates is a small trolley wheel, T', Fig. 153, borne by the supporting ring. The electric connections of the follow-up system are shown in Fig. 157. In this figure, A represents the armature and FF' represent the field coils of a reversing motor fastened to the spider. The two pairs of contactor plates XY and $X'Y'$ are attached to the opposite sides of the phantom element, and the two trolleys TT' are attached to opposite sides of the sensitive element. Since the motor is geared to the azimuth gear Z, Fig. 154, it is called the "azimuth motor." When the trolleys touch the contactor plates XX', as indicated in the diagram, current from B follows the course $BTXFAB$, and the motor armature together with the connected phantom element and system of contactor plates rotate in one direction. This rotation continues till the contactor plates YY' touch the trolleys TT'. Then the current follows the path $BT'Y'F'AB$. The direction of rotation of the azimuth motor and connected phantom element now is opposite that in the former case. The phantom element "hunts" back and forth through an angle of less than a degree about the sensitive element as mid-position. This oscillation results in elimination of lag in the indications of the sensitive element.

121. The Method of Damping. — If there were no damping, the north-seeking end of the gyro-axle would move in an elliptical orbit in a plane perpendicular to the meridian plane, Fig. 137. The amplitude of oscillation will be damped if either the vertical or the horizontal component be reduced. If both components be reduced, the damping will be greater. In the Sperry gyro-compass both components are reduced by an arrangement that causes the major axis of the elliptical path traced by the end of the precessing spin-axle to be inclined to the horizontal. The arrangement consists in connecting the mercury ballistic loosely to the gyro-casing by means of a pin that is slightly to one side of the vertical line through the center of mass of the gyro-casing and wheel.

As shown in Art. 107, if the mercury ballistic were connected to the gyro-case by a pin vertically below the center of the gyro-wheel, then the prolongation of the north-seeking end of the gyro-axle would trace on a vertical plane XZ, Fig. 158, a path which is an ellipse with major axis horizontal. Now we shall consider the effect of connecting the mercury ballistic to the gyro-case by a pin, e, to the east of the vertical through the center of the gyro. Suppose at some instant the spin-axle is horizontal and the north-seeking end is directed toward a point a, Fig. 158, east of north. Owing to the rotation of the earth, mercury flows from the more easterly reservoirs to the more westerly (Art. 107). This causes the phantom P to press against the pin e and so produces a torque about an axis oL, perpendicular to the line oe. The torque produces a precession about the line oe, inclined to the vertical at an angle θ. Since the axis of precession is inclined to the vertical, the prolongation of the spin-axle will cross the meridian at a point b'

nearer the horizontal than it would if the precession axis were vertical, Fig. 158. It will cross the horizontal at a point c' nearer the meridian than it would if the precession axis were vertical. It will cross the meridian again at a point d' nearer the horizontal than it would if the precession axis were vertical. In fact, the prolongation of the north-seeking end of the spin-axle traces on a vertical plane XZ, perpendicular to the meridian plane, a spiral that converges to a point on this plane very close to the intersection of the meridian and the horizontal plane through the center of mass of the gyro. When the compass is in northern latitudes, this point is slightly above the horizontal; when in southern latitudes, it is slightly below. The principal axis of the spiral is inclined to the horizontal plane and is fixed with reference to the precessing gyro-wheel. The degree of damping is controlled by the distance between the pin e and the vertical through the center of the gyro. The amplitude of each swing of the gyro-axle is about one-third that of the preceding swing.

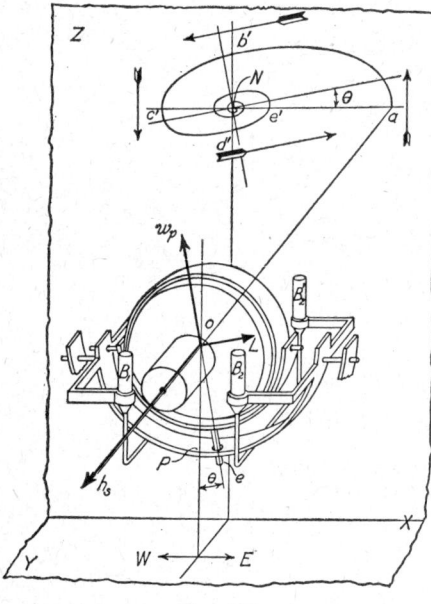

Fig. 158

In Fig. 158, the torque acting on the gyro-wheel is represented by L. This torque produces a precessional velocity represented by w_p, toward b'. If the mercury ballistic is maintained always in the same position relative to the gyro-axle, then the projection of the gyro-axle will continue to leave the elliptical path and trace a converging spiral having a major axis inclined to that of the ellipse. The constancy of the position of the mercury ballistic relative to the gyro-wheel is maintained by means of the follow-up system. This causes the ballistic to follow every movement in azimuth of the spin-axle. By means of the ballistic, supported by the phantom, and the eccentric connection with the gyro-case and mercury ballistic, the spin-axle will quickly settle to a position which is nearly horizontal and in the geographic meridian.

122. The Magnitude of the Latitude Error for Which Correction Must Be Made in the Sperry Gyro-Compass. — When the gyro-compass is at any latitude there is a vertical component of the earth's angular velocity w_e of the value, Fig. 159,

$$w_{ev} = w_e \sin \lambda$$

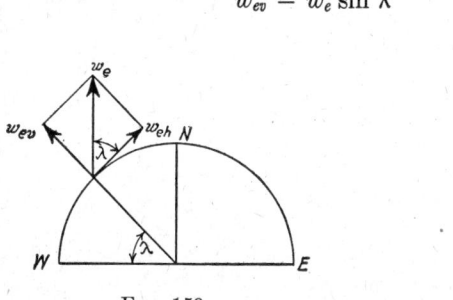

Fig. 159   Fig. 160

It follows that, in order that the spin-axle may remain in the meridian, the spin-axle must be caused to precess about a vertical axis with an angular velocity equal in magnitude to w_{ev} but in the opposite direction. This precession requires a torque about a horizontal axis. The requirement of a torque about a horizontal axis, and the additional requirement that for damping there must at the same time be a torque about the vertical axis, are fulfilled in the Sperry Mark V compass and later models by connecting the mercury ballistic to a point of the gyro-casing slightly to the east of the point vertically below the center of the gyro-wheel, as we have seen in Arts. 107 and 121. By this scheme, the applied torque is about an axis making a small angle θ to the horizontal, and the precessional velocity w_p thereby produced is about an axis inclined at the angle θ to the vertical.

The horizontal component of the torque produces a precessional velocity w_{pv} about a vertical axis, Fig. 160. The vertical component of the torque produces a precession w_{ph} about a horizontal axis perpendicular to the gyro-axle of the value

$$w_{ph} = w_{pv} \tan \theta$$

If the spin-axle of the compass is to remain in the meridian, the vertical component of the velocity of precession must equal the vertical component of the earth's rotation. That is,

$$w_{pv} = w_{ev}$$

Substituting in this equation the values above,

$$\frac{w_{ph}}{\tan \theta} = w_e \sin \lambda$$

$$w_{ph} = w_e \sin \lambda \tan \theta \tag{128}$$

In order that the gyro-compass may remain in equilibrium, the resting position of the gyro-axle, aa', Fig. 161, must make an angle δ_2 with the meridian plane NS such that the component w_{ph} of the precessional velocity of the spin-axle about a horizontal axis hh' perpendicular to the spin-axle, will equal the component of the earth's rotation with respect to the same axis. In Figs. 159 and 161, the horizontal component in the meridian plane of the angular velocity of the earth is represented by w_{eh}. Thus, we wish to have

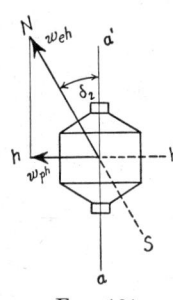

Fig. 161

$$w_{ph} = w_{eh} \sin \delta_2$$

From Fig. 159

$$w_{eh} = w_e \cos \lambda$$

From these last two equations and (128) we can eliminate w_{eh}, w_{ph} and w_e, and obtain

$$\sin \lambda \tan \theta = \cos \lambda \sin \delta_2$$

For all compasses designed for the same angle θ, $\tan \theta$ is a constant. If δ_2 be small, we may replace $\sin \delta_2$ by δ_2. Under these conditions, the correction of a Sperry compass due to latitude has the value

$$\delta_2 = \tan \theta \tan \lambda = b \tan \lambda \tag{129}$$

where b represents the constant $\tan \theta$. This correction must be added to or subtracted from the compass reading, depending upon whether θ is positive or negative and whether the compass is north or south of the equator. In northern latitudes the deflection is east of the meridian; in southern latitudes it is west of the meridian. At latitude 50°, the latitude error is about 2°.

123. Correction Mechanism for Velocity and Latitude Errors. — The compass card is surrounded by a coplanar ring called the "lubber ring" on which is engraved a line called the "lubber line" parallel to the keel of the ship. If the spin-axle is in the meridian plane, the angle between the N-S line of the compass card and the lubber line gives the direction of the ship. The cards of gyro-compasses are marked in degrees, from 0 to 360.

A person could tell the correct time from the reading of a clock known to be five minutes fast by subtracting from the observed reading the five-minute error. He could tell the correct time without the necessity of keeping the error in mind if the clock face were rotated through 30 degrees in the clockwise direction about the spindle which carries the two hands. This procedure might not be regarded as correcting the clock error but it would at any rate make the clock indicate the correct time. In a similar manner the meridian-steaming error and the latitude error of the Sperry gyro-compass are allowed for automatically by a shift of the lubber line.

The total deviation of the gyro-axle from the meridian due to the meridian-steaming error together with the latitude error is (119 and 129):

$$\delta_1 + \delta_2 = \frac{av \cos \theta}{\cos \lambda} \pm b \tan \lambda \qquad (130)$$

where a and b are easily determined constants, v is the speed of the ship in knots, θ is the course measured in degrees from the meridian and λ is the latitude. The first term in the right-hand member of (130) results in an eastward deflection of the north-seeking end of the spin-axle when the ship is steaming southward and a westward deflection when steaming northward. The second term results in an eastward deflection when the ship is in northern latitudes and a westward deflection when in southern latitudes.

The device used on the Sperry gyro-compass consists of two dials, a cam and a system of rods that maintain the proper shift of the lubber line after the dials have been set for the latitude and speed of the ship. Figure 162 is a diagram of the device that is used on compasses of the models designated Mark II and Mark V. The lubber ring X is turned relative to the compass card Y by the system connected to the pin Z. The course-corrector ring R is fastened rigidly to the phantom and consequently remains in constant relation to the gyro-axle of the compass. Its plane is inclined to the horizontal plane. The east-and-west diameter of the ring is horizontal. The higher side of the ring is directly above the north-seeking end of the gyro-axle of the compass. A roller C on one end of the rod AC engages in a groove in the edge of the ring. The end A of the rod AC and the two dials D_1 and

D_2 are fixed to the spider. When the course of the ship changes, the ring and the gyro-axle preserve their direction in space whereas the rod AC and all the remainder of the correction device turns with the ship. When the ship with the rod AC turns relative to the ring R, the roller C moves either up or down as it follows the groove in the ring. When the course of the ship is inclined to the

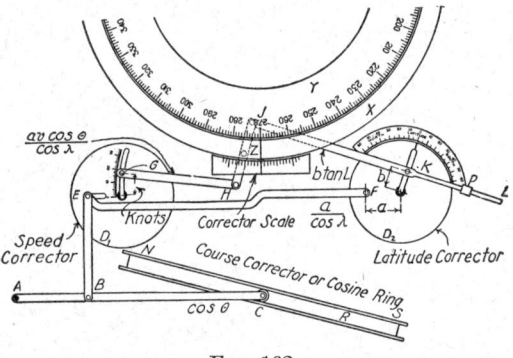

Fig. 162

meridian, the roller is displaced vertically from the horizontal plane through the center of the ring by an amount x which now will be determined. Figure 163 represents the speed-corrector ring R, in elevation and in plan, when the course of the ship makes an angle θ with the meridian. The inclination of the plane of the corrector ring to the horizontal is ϕ. From the figure:

$$x = d \tan \phi = r \cos \theta \tan \phi$$

Since $r \tan \phi$ is constant, the above equation shows that the vertical displacement x varies directly with the cosine of the course of the ship. This displacement produces a corresponding displacement of the pin E fastened to the dial D_1.

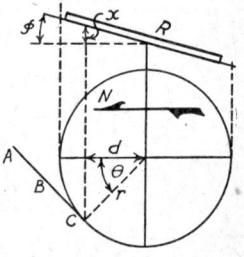

Fig. 163

A slot in the arc of a circle extends from the center to near the edge of the speed-corrector dial D_1. The radius of the arc equals the distance between the two pins at the ends of the rod GH. When the speed corrector dial is rotated from the position shown in the diagram, the rod GH and the pin Z are moved along. The amount of displacement of the lubber ring thereby produced depends upon the distance v of the pin G

THE SPERRY GYRO-COMPASS

from the center of the dial D_1. After the pin G has been clamped at the position corresponding to the speed v of the ship, the lubber ring will be displaced automatically by the amount required to allow for the quantity v in (130). Figure 162 is for a due westerly course. In this case there is zero correction for speed.

The pin E can move freely along the radial slit in the speed corrector dial. When the course of the ship is in the east-west line, the slot is horizontal as shown in the diagram. The position of the pin E in this slot is determined by the position of the pin F in the latitude-corrector dial D_2. The distance of the pin F from the center of the latitude corrector dial is made proportional to the constant a in (130). The length of the rod EF equals the distance between the centers of the two dials. When the latitude corrector dial is set for zero latitude, the pins E and F are on the straight line through the centers of the two dials. When the index is placed opposite the number representing the latitude of the ship, this arrangement makes allowance for the quantity $\dfrac{a}{\cos \lambda}$ in (130).

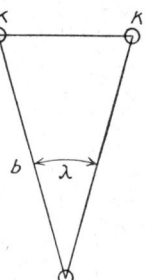

Fig. 164

One end of the rod JL can slide in a fixed guide P. This rod is fastened to the latitude-correction dial D_2 by means of a pin and clamp K that permit rotation of the dial. The distance of the pin from the center of the dial is made proportional to the constant b in (130). If while the end H of the rod HJ remains fixed, the index of the latitude correction dial be set at the number corresponding to the latitude λ of the ship, that is the clamp K moved to a new position K', then the edge of the lubber ring will be moved from the zero position through a distance proportional to $KK'[= b \tan \lambda]$, Fig. 164. Thus proper allowance would be made for the second term in the right member of (130).

The speed and latitude corrector used on the Sperry gyrocompass Mark VI and VIII is much simpler in design than that just described. Two blocks, xx', Fig. 165, are attached to the lubber ring. The other part of the device, NK, is attached to the spider. The two parts are connected by a stud Z projecting from a nut that can be moved back and forth by means of the knurled headed screw Y. The face of the nut is marked off into divisions representing latitudes. By setting the division corresponding to the latitude opposite the lubber line, the lubber line is shifted relative to the spider by an amount proportional to the quantity $b \tan \lambda$ (130).

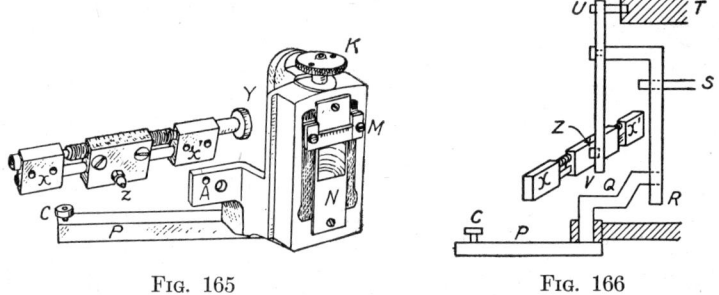

Fig. 165   Fig. 166

The course corrector cam consists of a circular groove in the under face of the azimuth gear, Z, Fig. 154. The groove is eccentric relative to the azimuth gear. In this groove fits a roller C on the end of a bent lever PQ, Figs. 165 and 166. The other end of the bent lever fits into the lower end of a vertical lever RT, Fig. 166, capable of rotation about a pin S. This pin can be moved up or down by means of a screw, K, Figs. 155 and 165, thereby changing the lever arm ST, Fig. 166. A displacement of the end T rotates the lever UV about a fixed pin U, thereby shifting the lubber ring attached to XX' by an amount which depends upon the lengths of the various lever arms. The lever PQ allows for the quantity $a \cos \theta$. The allowance for $\dfrac{v}{\cos \lambda}$ is made by a proper setting of the pin S. This setting is effected by adjusting the screw K, Fig. 165, until the correct latitude reading on the horizontal scale M is on the correct speed curve engraved on the face plate N, Figs. 156 and 165. The speed curves are obtained empirically.

By thus shifting the lubber line to the proper amount, the ship will point correctly but the compass will not. In taking bearings of heavenly bodies, or of points on the shore, it is necessary to know true directions. True directions can be obtained by means of a repeater compass electrically connected to a transmitting device mounted on the lubber ring of the master compass (Art. 127).

124. Avoidance of the Ballistic Deflection Error. — In the Sperry gyro-compass models designated Mark V, Mark VI, Mark VII and Mark VIII, the period of vibration of the sensitive element required to avoid the ballistic deflection error is secured by regulating the amount of mercury that flows from one side of the sensitive element to the other.

The mercury ballistics of the Sperry Mark VI and Mark VIII gyro-compasses have two separate reservoirs B_1, B_2, Figs. 152, 156 and 158, on the north side of the phantom element, connected by two tubes to correspondingly situated reservoirs on the south side. The mass of mercury that can pass from one side to the other is such that the ballistic error is zero when the ship is at latitude 40° and is not great at any latitude.

The Mark V gyro-compass has a single mercury reservoir on the north side of the phantom element and another on the south side. These reservoirs are divided into compartments and provided with valves which can be adjusted to give four different masses of active mercury. With one weight of active mercury, the torque that opposes tilting is such that the ballistic error is very small at all latitudes between 0° and 35°; with another weight, the error is very small at all latitudes from 30° to 50°; with another weight, the error is very small at all latitudes from 45° to 62°; with another weight, the error is small at all latitudes from 60° to 70°; in each case the ballistic error is zero at the mean latitude.

The period of the Sperry gyro-compass models designated Mark IX, Mark X and Mark XI is adjusted to the value proper to any given latitude by changing the lever arms of the weights of the active liquid in the mercury ballistic. This result is accomplished by moving four mercury reservoirs to the proper distance from the vertical axis through the center of mass of the suspended system. Each pair of reservoirs is fastened on opposite ends of a jointed tube similar to the jointed gas-light brackets sometimes used over work benches. By means of a worm and gear provided with a divided scale the reservoirs can be moved to the position proper to the known latitude. This device can be recognized in the Frontispiece. Observe the horizontal shaft carrying a worm at each end, on the right-hand side of the sensitive element. Each worm engages in a horizontal gear to which is rigidly attached a cylindrical reservoir. The reservoir attached to the gear that engages with the worm on the left-hand end of this horizontal shaft is only partially obscured by the gear. Observe the small tube joining this reservoir with another on the left-hand side of the sensitive element.

125. The Automatic Ballistic Damping Error Eliminator. — The Mark X gyro-compass is provided with an automatic damping eliminator which moves the pin e, Figs. 152 and 158, to a position vertically below the center of mass of the gyro-wheel when the ship has made a turn of as much as 15 degrees. Thereafter there is zero damping. The operation of the device is as follows: While the ship is making a turn, the compass binnacle with the attached spider element turns with respect to the phantom element and the gyro-axle. The large azimuth gear attached to the phantom element is in mesh with a small pinion attached to the spider element. The gear ratio is 3500 : 1. The pinion is connected to a shaft on which is mounted a group of three balls attached to three springs and a collar capable of sliding along the shaft. When the shaft and balls rotate, the balls fly outward thereby moving the collar along the shaft. When the ship is turned rapidly through an angle as great as 15 degrees, the sliding collar closes the electric circuit of a magnet which then moves the eccentric pin to a position vertically below the center of mass of the gyro-wheel. The damping now ceases and further ballistic damping error is prevented. When the turning of the ship ceases, the ball governor ceases to rotate, the electric circuit to the eliminator magnet is broken, the eccentric pin returns to its normal position and damping is renewed.

126. Avoidance of the Quadrantal or Rolling Error. — The sensitive element of the Sperry mercury ballistic gyro-compass is non-pendulous both when the mercury ballistic is in position and when it is detached. So long as the spin-axle is horizontal, all solid masses are balanced with respect to the horizontal axis. Forces due to rolling or pitching tend to accelerate these solid masses but they also will be balanced, and, consequently will produce zero deflection of the spin-axle. However, the meridian component of forces due to rolling or pitching will cause mercury to flow from one reservoir to the other, thereby causing the center of mass of the sensitive element to move to one side of the vertical line through the point of support. The oscillation of the mercury causes the sensitive element to act like a pendulum and to have a quadrantal error. By adding a small mass of metal to the top of each mercury reservoir, the sensitive element is made slightly anti-pendulous or top-heavy, thereby diminishing the quadrantal error produced at the end of each roll or pitch.

The dimensions of the apparatus are so selected that there is a phase difference of nearly one-quarter period between the oscillation of the sensitive element and the mercury from one reservoir to the other. As the residual slight deflections are not cumulative, they produce zero resultant deflection of the spin-axis.

127. The Repeater System. — The latitude error and the meridian-steaming error produce a deflection of the gyro-axle out of the meridian plane (Arts. 110 and 111). It follows that the card of the master compass does not give true directions relative to the meridian. When, however, the lubber line is shifted relative to the compass card through an angle equal to the latitude and meridian-steaming errors, the course of the ship relative to the

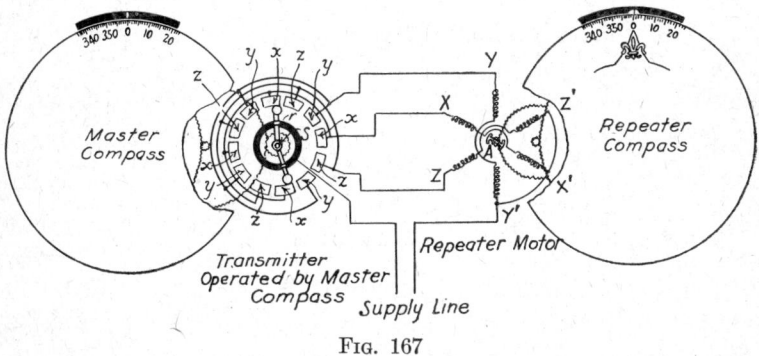

Fig. 167

meridian is properly indicated. The master compass gives true headings but not true bearings. The angle which any horizontal line through the compass makes with the meridian plane equals the angle between the given line and the gyro-axle plus the angle through which the lubber line is shifted from its zero position. The sum of these two angles is the bearing of any object on the given line. The addition of these two angles is made automatically by a repeater system consisting of a transmitter T, Figs. 155 and 156. The transmitter is electrically connected to repeater compasses, course recorder, automatic pilot, radio direction finder or other devices situated at convenient places on the ship.

The transmitter of the Mark VI and Mark VIII models is represented diagrammatically in the left side of Fig. 167. It consists of a collector ring r, twelve terminal blocks x, y, z, arranged in a circle concentric with the collector ring, and a contact-making transmitter arm capable of rotating about the central shaft. The transmitter arm is bent. The shaft s about which the arm rotates is at the elbow. On each end of the transmitter arm is a trolley which presses against the circle of contact blocks. At the elbow are brushes which press against the collector ring. The shaft s fastened to the transmitter arm is connected by a gear train to the azimuth gear of the compass. When the ship and the lubber ring turn with respect to the sensitive element, the transmitter arm rotates inside of the circle of terminal blocks.

Each repeater compass, course recorder or other auxiliary operated by the master compass has a repeater motor which is represented diagrammatically on the right side of Fig. 167. The motor has three pairs of poles and a soft iron armature A. When one pair of diametrically opposite poles is energized, the soft iron armature sets itself in line with that diameter, thereby rotating any compass card, course recorder or other auxiliary attached to the armature. When the transmitter arm rotates in the counter-clockwise direction, contacts are made in the sequence x, y, z, x, y, etc., and the repeater armature rotates in the same direction. When the transmitter arm rotates in the clockwise direction, contacts are made in the sequence x, z, y, x, z, etc., and the repeater armature rotates in the same direction.

With the transmitter arm in the position shown in the diagram, current goes from the direct current supply line to the collector ring r, to two x contact blocks, to the motor field coils XX' and back to the generator. The armature A sets itself in the direction XX'. If the transmitter arm rotates clockwise till one trolley is in contact with an x block while the other trolley is still in contact with a y block, then both the YY' field coil and the XX' field coil will be magnetized, and the armature A will set itself midway between them. Thus while the transmitter arm is making one revolution, the repeater armature makes one revolution in twelve steps.

The gear ratio between the repeater motor armature and the connected repeater compass card of the Mark VI and the Mark VIII Sperry gyro-compasses is 1 to 180. Consequently one revolution of the repeater motor armature rotates the connected repeater compass card two degrees. Therefore each electric impulse from the transmitter produces a rotation of the repeater compass card of one-twelfth of two degrees or ten minutes of angle. That is, a Mark V, Mark VI or Mark VIII gyro-compass repeater will indicate a change in the ship's course of five minutes of angle. The Mark X indicates a change in course of one-hundredth of a degree.

Since the transmitter is attached to the lubber ring, the corrections made for latitude and velocity by shifting the lubber ring are included in the indications transmitted to the repeaters.

§4. *The Brown Gyro-Compass*

128. Production of the Meridian-Seeking Torque. — The Brown gyro-compass is an instrument developed in 1916, of the pendulous type, with a single gyro-wheel of 4.25 lb. spinning at

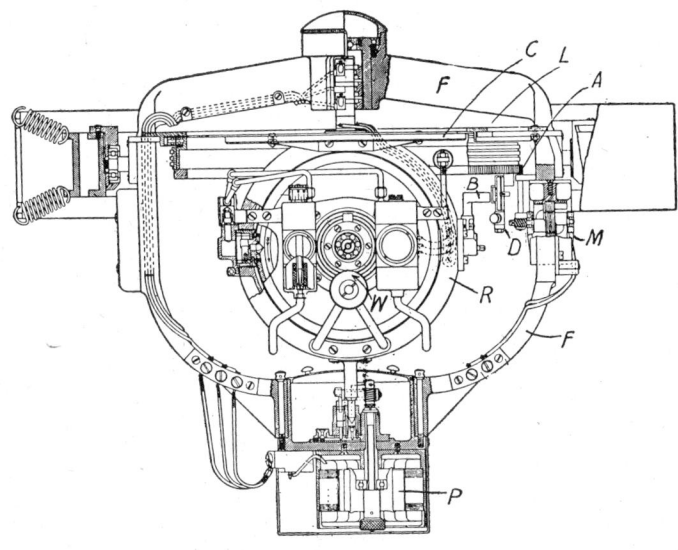

Fig. 168

about 15,000 revolutions per minute.* The master compass, removed from the binnacle, is shown in Figs. 168 and 174. The gyro-casing is carried by horizontal trunnions mounted in bearings on a ring R, Figs. 169 and 174, which always is nearly vertical. The vertical ring is capable of turning about a vertical shaft extending above and below the ring. The entire sensitive element is carried by an outer frame suspended from the gimbal rings of the binnacle not shown in the figures. The sensitive element does not rest on a solid bearing but upon a column of oil at high pressure under the lower shaft. Friction about a vertical axis is reduced to almost zero by rapid pulsations of oil pressure which move the sensitive element up and down through an amplitude of

* Made by S. G. Brown, Ltd., London, England.

about one-eighth inch some 180 times per minute. The compass card is attached to the vertical ring and the lubber ring is attached to the outer frame.

The unspinning gyro-wheel and casing constitute a non-pendulous system with respect to the horizontal trunnions connecting it to the vertical ring. The vertical ring and supporting frame constitute a pendulous system capable of oscillating about a horizontal axis. The required meridian-seeking torque is produced by displacements of oil between two connected "control" or "working bottles" C_1, C_2, Fig. 169, fastened to the gyro-casing G. The displacement of oil is due to an air-blast issuing from an upward directed nozzle Z fixed in the vertical supporting ring R. The rotation of the gyro-wheel produces a strong blower effect. The trunnions carrying the gyro-casing are tubular, and one of them conducts air to the nozzle Z. When the ring is vertical and the spin-axle is horizontal, this air-blast enters the two sides of the "air-box" A equally and produces equal pressures on the surfaces of the oil in the two control bottles. When the gyro-axle is tilted as in Fig. 169, oil is forced from the lower to the higher control bottle, thereby developing a torque on the gyro-system about an axis perpendicular to the plane of the diagram and in the direction represented by the symbol at L. If the direction of spin is in the direction represented by the arrow h_s, then the gyro-axle will precess about a vertical axis in the direction represented by the arrow w_p.

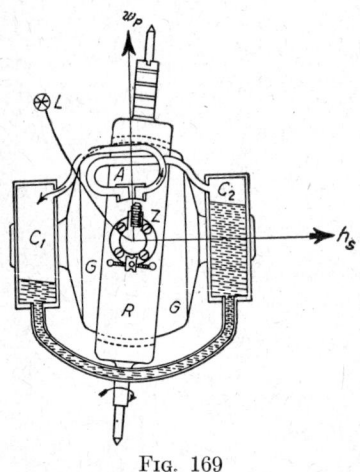

Fig. 169

Now consider the effect of the rotation of the earth on the direction of the spin-axle of a Brown gyro-compass. Imagine that we are above the northern hemisphere of the earth looking at a Brown gyro-compass at X (Fig. 170). Suppose that the spin-axle is horizontal, not in the meridian plane, and that the direction of spin is that represented by the arrow marked h_s. The pendulous frame and ring are vertical. Because of the rotation of the earth, the instrument will move into the position Y. The pendulous frame and ring continue to be vertical and the direction of the spin-axle continues in the same direction in space that it was when the instrument was at X. The resulting displacement of the gyro-casing with respect to the vertical frame causes an excess of air to enter the western control bottle and consequently an excess of oil to enter the eastern bottle. A torque is thereby produced about

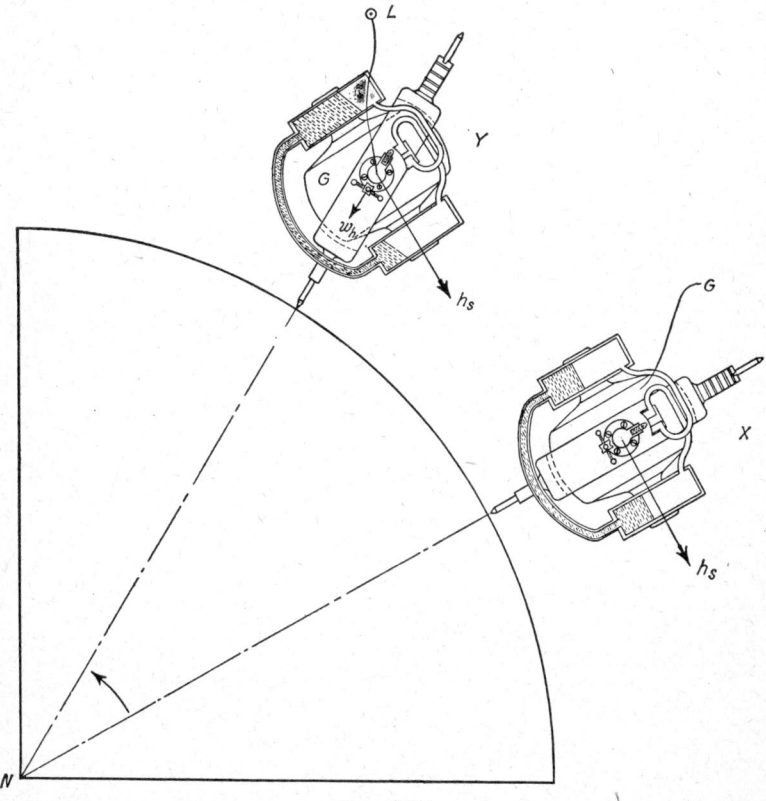

Fig. 170

an axis perpendicular to the spin-axis and in the direction represented by the symbol at L, Fig. 170. As the spin-axle moves toward parallelism with this torque-axis, it moves toward the meridian plane with the direction of spin in the direction of the rotation of the earth. The displacement of oil from one control bottle to the other causes this non-pendulous sensitive element to act as a pendulously mounted gyro. For this reason it is common to speak of the "degree of pendulousness" of the Brown sensitive element. As shown in Art. 106, the spin-axle crosses the meridian plane with the north-seeking end above the horizontal plane through the center of the gyro-wheel. It does not remain in that position, but slowly describes the surface of an elliptical cone having as axis a horizontal line in the meridian plane through the center of the gyro-wheel. The control bottles and air system constitute an air pressure liquid relay that causes the statically non-pendulous sensitive element to precess into the meridian plane with the direction of spin of the gyro in the direction of rotation of the earth.

129. The Method of Damping. — The oscillation of the spin-axle back and forth across the meridian plane is due to the tilting of the spin-axle from the horizontal. The oscillations can be damped by a torque about a horizontal axis opposing the torque produced by the pendulous vertical ring and supporting frame. This counter torque will be most effective in producing damping if, at every instant, it is proportional to the angular velocity of the sensitive element about the vertical axis.

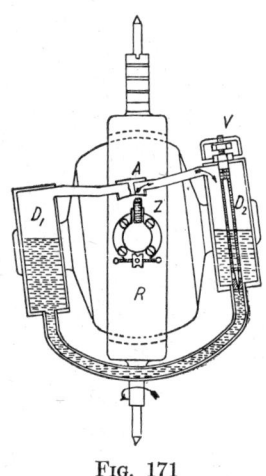

Fig. 171

In the Brown gyro-compass this result is accomplished by the displacement of a mass of oil back and forth between two metal bottles fastened to the opposite faces of the gyro-casing, beside the control bottles. The upper ends of these " damping bottles " are connected by small pipes to an air-box A above the upturned air-nozzle Z, Fig. 171. Note that the connections of the air-pipes from the control bottles to the air-box, Fig. 169, are not the same as the connections of the air-pipes from the damping bottles to the air-box, Fig. 171. A needle valve V controls the speed of flow of oil from one damping bottle to the other.

The operation of the damping device is as follows: Suppose that the spin-axle is horizontal and at its maximum angular displacement to the east of the meridian. The north-seeking end is at a, Fig. 137. The oil is at the same level in the control bottles, and also in the damping bottles. As the earth carries the gyro-compass from the position X to the position Y, Fig. 170, the spin-axle tilts, oil is forced from the lower control bottle into the upper, and the spin-axle turns toward the meridian. When the meridian plane is

THE BROWN GYRO-COMPASS

being crossed, the tilt of the spin-axle is maximum, the amount of oil in the upper control bottle is maximum, and the angular velocity of the spin-axle about a vertical axis is maximum.

Meantime, oil has been moving slowly from the upper damping bottle into the lower. So long as the spin-axle is nearly horizontal there is nearly the same amount of oil in the two damping bottles and consequently there is little opposition to the torque due to the control bottles. By the time the spin-axle is crossing the meridian plane with maximum speed, the difference in the amount of oil in the two damping bottles is nearly maximum and the opposition to the motion of the spin-axle is considerable. The transfer of oil from the upper to the lower damping bottle produces a diminution of the pendulousness of the sensitive element and, consequently, a diminution of the deflection of the gyro-axle.

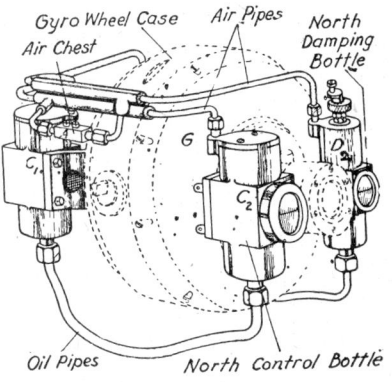

Fig. 172

The power acting on the sensitive element at any instant by the damping system equals the product of the angular velocity of the sensitive element and the torque opposing vibration at that instant. To produce damping, the phase difference between the torque and the angular velocity must be more than 90 degrees (Art. 88). The phase difference can be changed by regulating the flow of oil through the needle valve.

When the needle valve has been properly adjusted, oil flows so slowly from the upper damping bottle to the lower that the oil in these bottles has not become at the same level at the time when the spin-axle is horizontal. This difference in head causes the flow to continue in the same direction, notwithstanding the opposing air-blast, till the sensitive element is near the end of the swing. Throughout this time, the oil in the damping bottles produces a torque in opposition to the torque due to the pendulousness of the sensitive element. Hence, the deflections in azimuth of the spin-axle are diminished. After two or three vibrations, the spin-axle will come to a resting position. If no torque acts upon the sensitive element other than the ones already considered, the resting position of the spin-axle will be in the meridian plane.

We have described the Brown damping device as though there were two air-boxes and air-blast nozzles, one of each for the control bottles and one for the damping bottles. In fact, however, the control bottles and the damping bottles are piped to a single

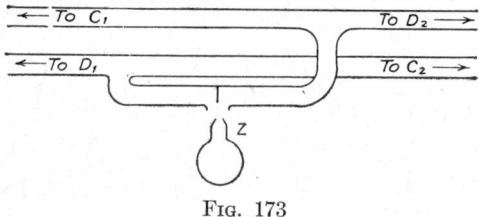

Fig. 173

air-box, Fig. 172. The air-box is shown in detail in Fig. 173. The pipes are joined to the north working and to the north damping bottles, and to the south working and south damping bottles as indicated.

130. Absence of Latitude Error. — In northern latitudes, the north-seeking end of the gyro-axle will have a resting position above the horizontal plane through the center of the gyro-wheel. In southern latitudes, it will have a resting position below the horizontal. The natural tilt of the spin-axle is the amount required to cause precession at the rate at which the meridian is turning, that is the amount required to cause the axle to precess into the resting position after it has been displaced from that position. If the torque due to the pendulousness of the gyro-compass be small, the spin-axle will have the same resting position that it would have if the pendulousness were greater, but the speed with which the spin-axle precesses will be slower.

The method employed to damp the vibrations of the Brown gyro-compass is equivalent to a diminution of the degree of pendulousness of the gyro-system while it is oscillating through the meridian plane. It does not oppose the tilt of the gyro-axle required to precess the gyro-axle to the normal resting position. It follows that the Brown gyro-compass is without latitude error (Art. 110).

131. The Meridian-Steaming Error. — Every gyro-compass is subject to an error depending upon the meridian component of the velocity of the ship (Art. 111). No means are provided for correcting the indication of the Brown master gyro-compass, but the meridian-steaming error is allowed for in the repeater compasses

THE BROWN GYRO-COMPASS

by an eccentric mounting of the repeater cards. The proper degree of eccentricity in the mounting of the repeater cards is made by setting dials for the given latitude, speed, and course.

132. The Repeater System. — The pendulous system of the Brown gyro-compass, Type A, is shown in Figs. 168 and 174. The gyro-wheel, gyro-casing with the attached two control bottles and the two damping bottles, the vertical supporting ring R and the attached compass card C, constitute the sensitive element. The outer frame F carrying the lubber ring L and the motor-driven oil pump P hangs on the Cardan rings of the binnacle and turns with the ship.

The angle between the lubber line and the north-south line of the compass card is to be transmitted to and indicated by the repeater compasses, course recorder and other repeater devices. A horizontal gear ring, A, Fig. 168, of about the same diameter as the compass card, is mounted on the outer frame concentric with the lubber ring.

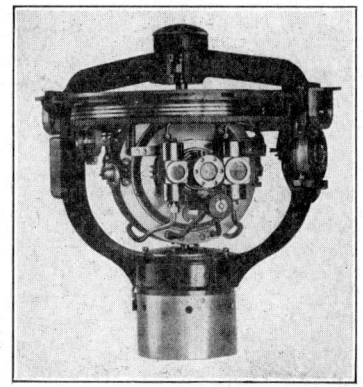

Fig. 174

The angle indicated by the repeaters is to be the angle between the lubber line and an index line on the gear ring when this index line is parallel to the north-south line of the compass card. The gear ring A can be rotated about a vertical axis by a reversible motor M fastened to the outer frame. Frictionless contact between the sensitive element and the gear ring is effected by an air-blast from the gyro-casing against two plungers of a pneumatic electric contact maker.

The gyro-casing is supported by two hollow trunnions carried by knife-edges on the vertical ring. An air-blast produced by the rotating gyro-wheel passes through one of these trunnions, emerges from a nozzle fixed to the vertical ring and enters the air box connected to the control and to the damping bottles. Another air-blast traverses the other trunnion, emerges from a second nozzle B attached to the vertical ring and blows against two adjacent plungers of the pneumatic electric contact maker D fastened to the lower face of the follow-up gear-ring. The two plungers are connected to a rocking arm that makes an electric contact causing the reversible motor to rotate in one direction when one plunger is pushed in, and in the reverse direction when the other plunger is pushed in. If the ship is headed in the direction of the spin-axle, the air-blast strikes the two plungers equally and no electric contact is made. As soon as the ship begins to turn, the supporting frame F turns relative to the sensitive element, thereby moving the two plungers relative to the air-blast nozzle B. One plunger is pushed more strongly than the other, the reversible motor starts and rotates the follow-up gear till the two plungers are pushed equally. The angle through which the follow-up gear-ring has been rotated is the angle the ship's head is turned from the north-south line. The follow-up gear is electrically connected to step-by-step motors which rotate the cards of the repeaters through the same angle that the follow-up gear has been rotated.

133. The Ballistic Deflection Error. — The resting position of the spin-axle of a compass depends upon the latitude and the meridian component of the velocity of the ship (Art. 111). When the velocity of the ship changes either in direction or in magnitude, the spin-axle is deflected from its resting position and oscillates about a new resting position. The magnitude of the so-called ballistic deflection depends upon the latitude, the meridian component of the linear acceleration of the ship's velocity, and upon the period of vibration of the gyro-axle back and forth across the meridian plane.

In Art. 113, it is shown that if the period of vibration of the sensitive element is of the proper value, the ballistic deflection error will be zero and the spin-axle will move without oscillation into the resting position proper for the new velocity of the ship. As so much sailing is done at latitudes about 40°, it is common practice to adjust the period of the Brown compass for zero ballistic deflection error at this latitude.

134. Prevention of the Quadrantal or Rolling Error. — While a ship is rolling or pitching, a jerk is imparted to the pendulous system at the end of each roll or pitch. It is shown in Art. 115 that these jerks produce zero deflection of the gyro-axle when the ship is headed on a cardinal course, north, east, south, or west, but that an error is produced when the ship is on an intercardinal or quadrantal course, unless means be taken to prevent it. Any error in the indication of a gyro-compass that might be produced by these jerks can be avoided by delaying the action of the jerks on the sensitive element till the axis of suspension is vertical.

In the Brown gyro-compass this result is produced by motion of the oil in the control bottles.

Owing to the rotation of the earth, the spin-axle of the compass tilts out of the horizontal plane. Oil is forced into the control bottle on the upper side of the gyro-case. As the earth rotates on its axis so very slowly, the flow of oil due to this cause ceases as soon as the tilting of the gyro-axle ceases.

When the ship rolls, the gyro-compass is forced into oscillation about the fore-and-aft axis of the ship with a period equal to that of the roll. The mean angular velocity of the compass relative to the spin-axle is some two thousand times the mean angular velocity of the earth about its axis. Now, the velocity of oil from one control bottle to the other is so great that the kinetic energy of the oil causes the flow to continue after the tilting of the spin-axle has ceased. Hence, the oscillation of the oil from one control bottle to the other lags behind the oscillation of the compass and behind the oscillation of the ship. The lag is made such that, when the ship is at its maximum inclination to the vertical, the oil is at the same level in the two control bottles; when the ship is upright and the air-blast is equally divided at the two compartments of the air-box, the oil is at its maximum height in the control bottle on the side of the gyro-wheel from which the ship has just righted itself. Thus, the jerks imparted to the compass at the end of a roll do not affect the gyro-wheel till the ship is on even keel, a quarter period later.

If the ship is heading on a quadrantal course, Fig. 149, the jerks imparted to the compass when the ship is at the end of a roll are not received by the sensitive element of the Brown gyro-compass till the axle of the vertical supporting ring is vertical. Hence, the linear motion of the north-seeking end of the spin-axle produced by these jerks can have no vertical component. There will be zero deflection from the meridian due to the rolling (Art. 117, *a*).

If the moment of inertia of an oscillating pendulous system is not the same with respect to all horizontal axes through the point of support, the system will tend to rotate about a vertical axis till the moment of inertia of the system is maximum with respect to the axis about which it is oscillating (Art. 14). A gyro-compass having unequal moments of inertia with respect to different horizontal axes through the point of support would exhibit quadrantal error. The moment of inertia of the Brown gyro-compass is made of the same value with respect to all horizontal axes through the point of support by attaching to each side of the vertical supporting ring a compensating mass W, Fig. 174.

§5. *The Anschütz Gyro-Compass*

135. The Sensitive Element of the Model of 1926. — The first seaworthy gyro-compass was produced in 1908 by Dr. Hermann Anschütz at Kiel, Germany. The model of 1926 has two gyro-wheels each with the spin-axle at an angle of about 30 degrees to the meridian. Each gyro-casing is capable of a slight rotation about a vertical axis. The two are connected by geared arcs so that, when the sensitive element is in equilibrium, the angles are equal between the meridian and the two gyro-axles.

Until 1926 the sensitive element was supported by a float in a vessel of mercury. When this method was used, the turning of

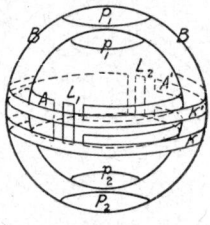
Fig. 175

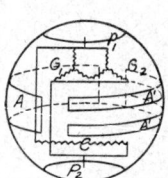
Fig. 176

the sensitive element about a vertical axis was opposed by a small but unconstant torque due to the surface tension at the mercury-air-metal line of contact. This slight drag is avoided in the model of 1926 by the very ingenious scheme of enclosing the gyro-wheels and damping device in a hermetically sealed spherical globe that is kept poised within an outer spherical shell filled with an electrolyte consisting of a mixture of acidulated water and glycerine. This globe with the enclosed apparatus constitutes the sensitive element.* The sensitive element is slightly pendulous.

The gyro-globe is about ten inches in diameter, and the outer spherical shell has an inside diameter of about ten and one-half inches. Both the globe and the outer spherical shell are made of thin metal. The outside of the gyro-globe and the inside of the spherical shell are covered by a thin layer of hard rubber vulcanized on the metal shells. The gyro-globe has round electrodes, p_1, p_2, Fig. 175, at the upper and lower poles, and also an equatorial

* U. S. Patent. Anschütz, No. 1589039, 1926.

electrode A, A'. This equatorial electrode is divided by two vertical gaps. One of these halves is further divided into two by a horizontal gap. All three parts are electrically connected. The spherical shell is provided with two polar electrodes P_1, P_2, and two ring-shaped electrodes K, K'. The electrodes consist of thin layers of hard rubber charged with carbon and vulcanized on the metal globes. All electrodes are inlaid in the insulating hard rubber coatings of the outside of the gyro-globe, or the inside of the spherical shell, so as to be flush with the spherical surfaces.

The volume of the gyro-globe is such that the weight of the liquid displaced by it is nearly equal to the weight of the globe. Any tendency of the gyro-globe to sink or to move laterally from the central position within the surrounding sphere is prevented by the magnetic force of repulsion developed by the interaction of the magnetic field about an alternate current carrying coil C, Figs. 176 and 177, within the globe, and the magnetic field of eddy currents induced in a conducting saucer-shaped electrode P_2 forming part of the lower side of the outer spherical shell. The centralizing coil produces a conical repelling field directed toward the center of the gyro-globe.

The centralizing coil C and the three-phase motor of the two gyros G_1, G_2, Fig. 177, are joined to the equatorial electrode A and to the two polar electrodes, as indicated in Fig. 176. A three-phase current passes from the electrodes P_1, P_2 and K, K' on the inside of the outer spherical shell, through the thin layer of electrolyte, to the corresponding electrodes p_1, p_2, and A, A' on the outside of the gyro-globe, Fig. 175.

Lubrication of the moving parts within the gyro-globe is effected by wicks dipping into a pool of oil in the bottom of the globe, x, x', Fig. 177a.

The gyro-globe is exhausted of air and then filled with dry hydrogen at atmospheric pressure. The use of hydrogen instead of air results in several advantages: (a) the oil required for the operation of the apparatus within the globe suffers no chemical change even during months of continuous service; (b) since windage loss is proportional to the density of the surrounding gas, the windage loss with hydrogen is about one-fourteenth that which would occur if the globe were filled with air at the same pressure; (c) since the thermal conductivity of hydrogen is about seven times that of air and the diffusivity is about four times that of air, it follows that hydrogen displaces air in any porous insulating medium and there-

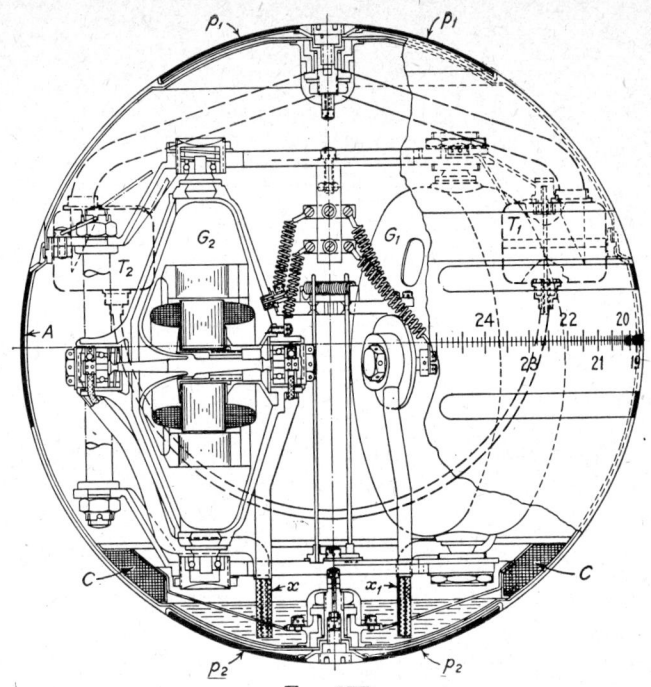

Fig. 177a

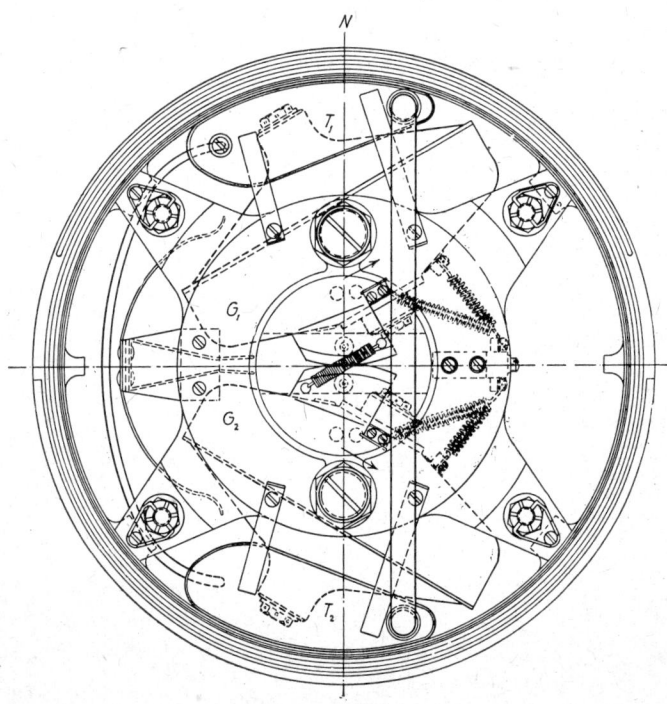

Fig. 177b

by increases the thermal conductivity of the porous material; (d) the high thermal conductivity of hydrogen improves the rate of transfer of heat across the film of contact; (e) since hydrogen displaces the oxygen that otherwise would be in the pores of the insulating material, damage of the insulation due to corona discharge is prevented.

136. The Supporting System. — The spherical shell B containing the sensitive element A is hung by six arms, DD, Figs. 178 and 179, to a vertical spindle S supported by ball bearings on the top of a bowl E which in turn is carried by the gimbal rings of the

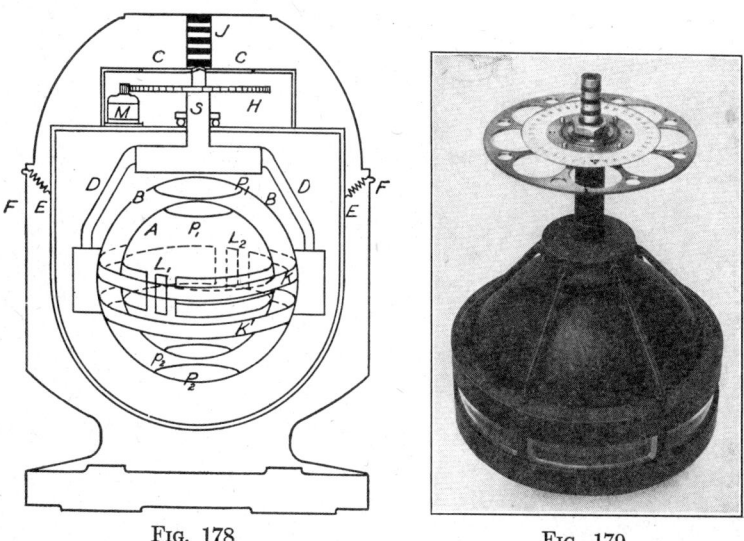

Fig. 178   Fig. 179

binnacle F. Attached to the vertical spindle are a horizontal gear H connected by a pinion to a reversing motor M, the compass card C, and five collector rings J which serve to connect electrically the master compass to the outside part of the equipment.

Wires from the collector rings extend down the supporting arms DD to the outer spherical shell B. One collector ring is connected to the two equatorial electrodes KK', another is connected to the polar electrode P_1, and another to the polar electrode P_2. The other two collector rings are connected to two electrodes L_1, L_2 attached to the spherical shell midway between the two equatorial electrodes KK'. The electrodes L_1 and L_2 are called the "follow-up" electrodes.

A top view of the compass removed from the binnacle is shown in Fig. 180. A group of springs connecting the compass to one of the gimbal rings, and a rod of felt between the springs and the compass, serve to reduce vibration. The thermometer at the left, and the two tubes that extend from the top plate of the compass to the trunnion in the foreground, are parts of the system that controls the temperature of the liquid in the outer containing vessel.

Fig. 180

137. Damping. — Two different methods are used on different models of the Anschütz gyro-compass for producing damping of the vibrations of the sensitive element. In the first method to be described, an annular tube is fastened, with the plane of the ring horizontal, to the frame that carries the gyros. The tube is partly filled with oil. If the sensitive element tilts, oil flows from one part of the tube to the part diametrically opposite. The flow is retarded by a set of partitions across the bore of the tube, each partition being pierced by an aperture of predetermined size. The operation of this damping system is as follows. When the meridian-seeking axis of the sensitive element moves east of the meridian, the north point of the compass card will rise till the gravitational torque due to the tilted pendulous system precesses the axis back to the meridian (Art. 106). The north point of the compass card continues to move after traversing the meridian, decreasing its upward tilt as it moves west of the meridian. After reaching a maximum displacement west it moves eastward and dips below the horizontal. After traversing the meridian in the eastward direction, the dip decreases and becomes zero when the maximum deflection eastward is produced.

While the sensitive element is tilting up and down the oil in the damping device moves back and forth. While the north-seeking end of the sensitive element is above the horizontal, that is while the north point of the compass card is moving westward, there is an excess of oil in the south side of the damping device. Similarly, there is an excess of oil in the north side of the damping device while the north point of the compass card is moving eastward. The torque produced by the ununiform distribution of oil in the damping device is, at all times, in the direction opposite the gravitational torque due to the tilted pendulous sensitive element. Thus the tendency to precess, due to the tilting of the oscillating sensitive element is diminished at every instant by a torque due to the displaced oil in the damping device. This torque does not oppose precession but reduces the effective pendulousness of the sensitive element.

In the second method, the oscillations of the sensitive element are damped by a pair of tiny Frahm anti-roll tanks, T_1, T_2, Fig. 177, fastened to the inner surface of the gyro-globe. Consider a gyro-wheel and casing to which are attached two connected oil tanks forming a pendulous system capable of oscillation about an axle through x, Fig. 181. The reservoirs are half filled with oil. If the connecting tubes were stopped up or if they were sufficiently large in diameter, the device would produce no damping of the vibrations of the pendulous system. If, however, there be a considerable opposition to the flow of oil, then when one reservoir is higher than the other, oil will flow slowly from the higher to the lower reservoir. By the time that the pendulous system is vertical there will be an excess of oil in the reservoir that was the lower, Fig. 182. Later, there will be a flow in the opposite direction, followed by a flow back and forth at a definite period. The period, as well as the phase difference between the oscillation of the oil and of the pendulous system, can be regulated by an adjustment of a valve in either of the tubes that connect the two reservoirs.

If the period of flow back and forth equals the period of vibration of the pendulous system, and if there is a phase difference of a quarter period between the motion of the oil and the motion of the pendulous system, the following effects will occur. When the axle of the spinning gyro-wheel is in the meridian plane, the north-seeking end of the axle will be higher than the south-seeking end and will be moving westward (Art. 106). At some such instant, there will be equal amounts of oil in the two reservoirs, Fig. 181. Since oil is flowing from the upper reservoir to the lower, the distance between the center of gravity of the pendulous system and the vertical line through its axis of vibration is being diminished. Hence the torque urging the pendulous system toward the equilibrium position is being diminished faster than it would be if the oil were not flowing. By the time the north-seeking end of the spin-axle has reached its maximum displacement toward the west and the spin-axle is horizontal, the oil has reached its maximum height in the southern reservoir, Fig. 182.

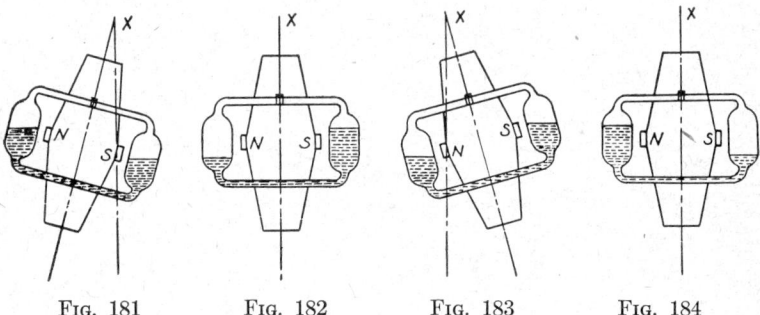

Fig. 181   Fig. 182   Fig. 183   Fig. 184

At the end of another quarter vibration of the pendulous system, Fig. 183, the spin-axle is in the meridian plane, the north-seeking end is dipping below the horizontal and the quantities of oil in the two reservoirs are equal. During this motion, the oil in the reservoirs diminishes the pendulous effect of the pendulous system and the spin-axle dips more slowly than it would have done if the oil were not flowing. By the time that the spin-axle again has become horizontal, Fig. 184, and its north-seeking end has reached its maximum displacement toward the east, the oil has risen in the northern reservoir to its maximum height. During this motion, the torque urging the pendulous system toward the equilibrium position is being diminished faster than it would be if the oil were not flowing. Consequently the amplitude of swing toward the east across the meridian plane is less than it would be if the oil were not flowing.

It is left as an exercise for the student to show that when the flow of liquid is adjusted as above specified, then the periodic torque acting on the gyro due to the surging liquid is in nearly opposite phase to the angular velocity of the tilting spin-axle.

This is the condition of nearly maximum damping (Arts. 25 and 27).

138. The Meridian-Steaming Error. — The meridian-steaming or north-steaming error occurs whenever a gyro-compass is being carried with a velocity which has a meridian component, that is whenever the ship is steaming on any course except directly east or west.

The magnitude of this error depends upon the course, the speed, and the latitude. As it does not depend upon the design of the instrument, it cannot be avoided. Its magnitude under various conditions can be computed and allowed for. A table of values of the meridian-steaming error is attached to the Anschütz compass. The errors may be taken account of by moving the lubber ring of the master compass and of the repeaters through the angle corresponding to the known course, speed, and latitude.

In northern latitudes the north-seeking end of the axle of a gyro-compass should tilt upward, and in southern latitudes it should tilt downward, through just the proper angle to cause the axle to precess toward the meridian plane at the rate required to keep the axle in the meridian plane. Any opposition to the proper tilting of the gyro-axle produced by damping will cause the rate of precession to be less than that required to keep the gyro-axle in the meridian plane. The settling position will be at an angle to the meridian plane called the latitude error.

The methods of damping the vibrations of the sensitive element of the Anschütz gyro-compass of 1926 do not reduce the tilt required, at the given latitude, to cause the axis of the sensitive element to precess into the meridian plane. In fact, the tilt must be greater than it would need to be if the damping device were not used, before the reduced effective pendulousness of the sensitive element will generate the necessary rate of precession at the particular latitude. There is no torque opposing the precession. Consequently, any latitude error is avoided (Art. 110).

139. Prevention of the Ballistic Deflection Error. — When a ship is steaming in any direction except on an east-west heading, the gyro-compass is deflected from the meridian plane by an angle called the meridian-steaming or north-steaming error (Art. 111). The resting position of the north-south line of the sensitive system depends upon the latitude and the meridian component of the velocity of the ship. If the velocity of the ship be changed either in direction or in magnitude, the north-south line of the sensitive system may not move at once to the resting position proper to the new velocity but may oscillate for some time. The angle between the new resting position and the resting position corresponding to the final velocity is called the ballistic deflection. In Art. 113 it is shown that a gyro-compass will be without ballistic deflection error at the equator if the undamped period has a value given by (126). In the same Article it is shown that if the compass be moved to any other latitude, the north-south line of the sensitive element will move without oscillation to the resting position proper to the final velocity of the ship so long as the quantity

$$\frac{K_s w_s}{mx \cos \lambda} \, [\, = Rw_e]$$

is maintained constant.

In the gyro-compasses already described, the period is adjusted to the proper value by regulating the pendulousness mx of the sensitive element. In the Anschütz and in the Arma gyro-compasses, the desired constancy of this quantity is maintained by varying the spin-velocity w_s of the gyro as the cosine of the latitude varies. Each gyro of the Anschütz compass is the rotor of a three-phase motor operated by a current of 330 cycles per second. By varying the inductance of two of the motor windings, the speed can be changed as gradually as may be desired from about 17,000 to about 30,000 revolutions per minute. By setting a dial for the given latitude, the gyros will have the proper angular momentum to produce the required period of vibration of the sensitive element. Details of a similar device used on the Arma gyro-compass as well as the means provided to prevent this change in spin-velocity affecting appreciably the direction of the spin-axle are given in Art. 145.

140. Avoidance of the Quadrantal Rolling Error. — Suppose that a ship on an intercardinal course carries a pendulous gyro-compass. When the ship is on even keel, the center of mass of the sensitive element of the compass is at C, Fig. 185. When the ship rolls, the center of mass of the sensitive element rotates readily about a north-south axis but not about an east-west axis. When the ship is at the end of a roll to port, the center of mass of the sensitive element has rotated to C' and is acted upon by force F'. When the ship is at the end of a roll to starboard, the center of mass of the sensitive element has rotated to C'' and is acted upon by a force F'''. The center of mass moves back and forth along a line perpendicular to the spin-axle.

The meridian components of these athwartship forces, produced by the shifting of the center of mass along the east-west line, produce a tilt of the spin-axle and also an azimuthal deflection when the ship is on an intercardinal heading (Arts. 115 and 117). This rolling, quadrantal or intercardinal error will be reduced to zero if there be no shifting of the center of mass of the pendulous sensitive element when the ship rolls. There will be only a very small back-and-forth displacement of this center of mass when the period of vibration of the sensitive element east and west is so much greater than the period of the ship's roll that the amplitude of the vibration forced upon the sensitive element is small.

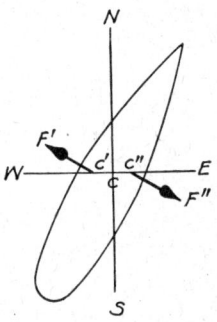

Fig. 185

In the Anschütz gyro-compass, the required long period of vibration of the sensitive element about a horizontal meridian axis is effected by having the sensitive element comprise two gyros with the spin-axles always nearly horizontal. The principle involved will now be described. Each gyro is capable of precessing through a limited angle about a vertical axis. The plane of the axis of precession is in the normal north-south plane of the sensitive element. The two gyro-casings are connected by cams and springs so that, at all times, the two spin-axles make equal angles with the vertical plane through the precession axis, Fig. 186. The angle between the spin-axles is always about 60 degrees.

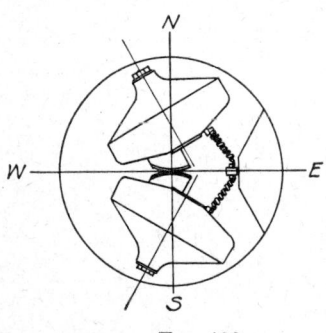

Fig. 186

If a change be made in the direction of the spin-axle of a gyro free to precess, a gyroscopic torque opposes the change. This opposing torque is proportional to the component of the angular momentum of the gyro with respect to an axis perpendicular to the axis about which the spin-axle is turned. With two gyros each of angular momentum h_s and the spin-axes horizontal and inclined 30 degrees to the meridian, there is a resultant angular momentum of the system with respect to the north-south line, of the value $2 h_s \cos 30° = 1.73 h_s$, Fig. 187. Along an east-west line, the angular momentum of the western gyro is $h_s \sin 30°$, while that of the eastern gyro is $-h_s \sin 30°$, the plus and minus signs referring to the directions of spin. The opposition to roll would be the same if the direction of spin of both gyros were reversed. Thus the opposition to roll about a horizontal north-south axis is proportional to $2 h_s \sin 30° = h_s$, and the opposition about an east-west line is proportional to $1.73 h_s$. These torques are so great that when the system is given jerks by the rolling or pitching of the ship, the period of the vibrations of the sensitive element thereby produced is from five to ten times as great as the period of the succession of jerks. This great difference of period precludes forced resonant vibration of the gyro-element in tune with the vibration of the ship. The jerks imparted to the pendulous system by the rolling or pitching of the ship produce a slight horizontal oscillation, but a negligible tilting of the line from the point of support to the center of mass of the pendulous system. Hence there is negligible quadrantal error. The centralizing device also opposes any rocking motion of the sensitive element.

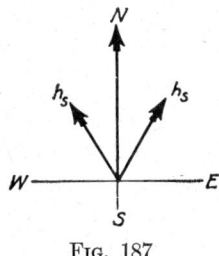

Fig. 187

141. The Follow-Up Repeater System. — As the sensitive element does not touch the supporting system which carries the compass card, a follow-up device must be employed to maintain the compass card in constant relation to the sensitive element. The scheme employed comprises a reversing motor, M, Fig. 178, pinioned to a horizontal gear H fastened to the supporting system. The method of controlling the motor will be described by the aid of Figs. 188 and 189, which represent the gyro-globe and spherical shell in elevation, and in plan, respectively. In Fig. 188, the two electrodes KK' are shown as though they were separated by a greater distance than the two electrodes A', A'. In the actual apparatus, this is not the case.

When the sensitive element with spinning gyro-wheels is in equilibrium and the north-south line of the compass card is in the meridian, the two " follow-up " electrodes, L_1, L_2, on the inside of the spherical shell, face the centers of the corresponding vertical gaps between the equatorial bands on the gyro-globe. The three lines X, Y, Z, of a three-phase current circuit, are joined,

THE ANSCHÜTZ GYRO-COMPASS

respectively, to the pair of equatorial electrodes on the inside of the outer spherical shell, to the upper polar electrode P_1 and to the lower polar electrode P_2. The broad equatorial electrode A on the outside of the gyro-sphere and the two polar electrodes p_1, p_2 are joined to the three terminals of the two three-phase gyro-motors. The broad and the two narrow electrodes on the gyro-sphere are connected together.

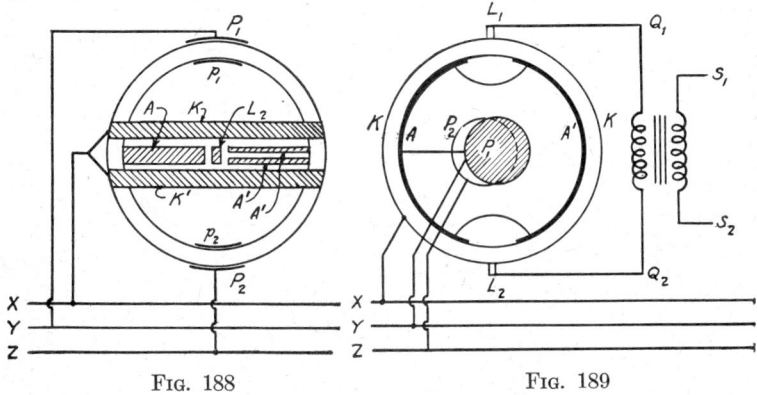

Fig. 188   Fig. 189

At the instant when X is at a higher potential than Y and Z, current from X goes to KK'. There it divides, a part crosses the electrolyte to A, traverses the gyro-motors to p_1, the electrolyte to P_1 and thence goes to the line Y. Meanwhile the other part crosses the electrolyte to A', passes to A, traverses the gyro-motor to p_1, the electrolyte to P_1 and thence to the line Y. These two paths are of different resistance.

With the spherical shell in some such position relative to the gyro-globe as that indicated in Figs. 188 and 189, the electromotive forces at L_1 and L_2 will be in the same phase. When the spherical shell becomes slightly turned from this position, by the turning of the ship for example, the electromotive forces at the follow-up electrodes L_1 and L_2 will be in different phase. The difference in phase will be reversed when the deflection of the spherical shell from the equilibrium position is reversed in direction. Thus, with any given phase difference between the lines X and Y, the wire Q_1Q_2 will be traversed by a current which in magnitude and in direction depends upon the angular displacement, from the equilibrium position, of the spherical shell relative to the gyro-globe. The variations of this current are caused to produce greater variations in the current of a three-electrode tube amplifying circuit.

236   NAVIGATIONAL COMPASSES

The terminals of the amplifying circuit and the line Z, Fig. 189, are joined to the terminals of a three-phase reversing motor M, Fig. 178.* By these means, the hollow sphere is turned till the follow-up electrodes are in the equilibrium position, that is, till the north-south line of the compass card is in the geographic meridian. The same lines operate repeater compasses, a course recorder, an automatic pilot, or other repeater devices.

§6. *The Arma Gyro-Compass*

142. The Sensitive Element. — The Mark IV Model 1 gyro-compass made by the Arma Engineering Company of Brooklyn, New York, Fig. 190, has two gyros capable of precession about

Fig. 190

vertical axes. The two gyro-casings are held by a bell crank and springs so that the angle between the spin-axles is kept constant, Fig. 191. When the sensitive element is in the resting position, the spin-axles are inclined about 40 degrees from the meridian.

The gyros spin at about 12,000 revolutions per minute in an atmosphere of helium under low pressure. The advantages in the use of helium are the same as in the use of hydrogen (Art. 135).

* U. S. Patent. Anschütz, No. 1586233, 1926. "The New Anschütz," published by the Nederlandsche Technische Handel Maatschappy "Giro," The Hague, Holland.

THE ARMA GYRO-COMPASS

The frame that supports the gyro-casings also carries two connected oil tanks, D_1, D_2, Fig. 191. One of these tanks is on the north and the other is on the south side of the frame when the element is in the resting position. The motion of oil back and forth from one tank to the other quickly damps vibration of the sensitive element (Art. 137).

Figure 192 represents the two principal units of the master compass pulled apart vertically. The gyros and attachments are

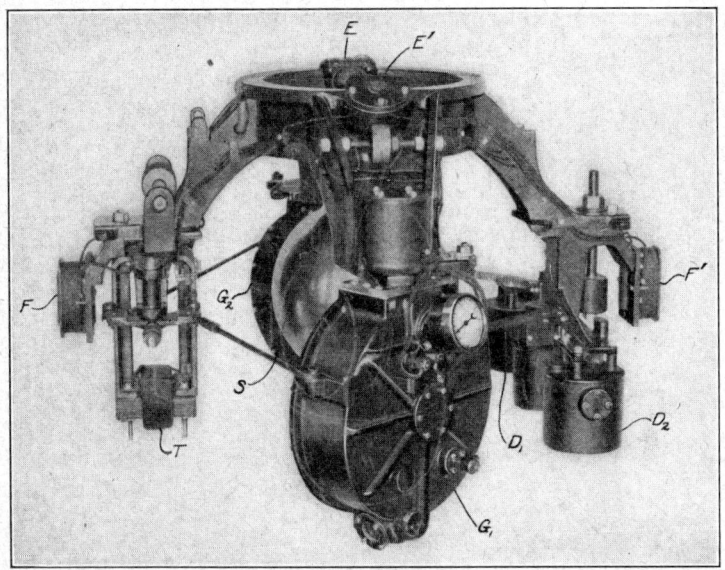

Fig. 191

supported by a hollow metal globe, S, that floats in a bowl of mercury, H. The globe together with the two gyros G_1 and G_2, with the entire supported system, constitute the sensitive element. Any change in friction that might be produced by variations in the surface tension at the line of contact of mercury, air and metal is prevented by a continuous vibration of the mercury maintained by a small motor fastened to the bowl and to the binnacle.

The compass card A, a horizontal gear and horizontal arm PP' are fastened rigidly to a vertical tubular shaft R. This system constitutes the phantom element. The phantom is carried by the compass frame. The latter is hung by springs from the inner gimbal rings of the binnacle. The phantom element is capable of being rotated about a vertical axis by a follow-up motor M. The tubular shaft R carries a group of slip rings against which press brushes for transmitting currents to the various electric devices attached to the phantom and to the sensitive element.

Three of the electric conductors for the operation of the sensitive element are within the tubular shaft projecting downward from the phantom and terminating in three pins near the center of the supporting globe. One pin is central and rests in a conical

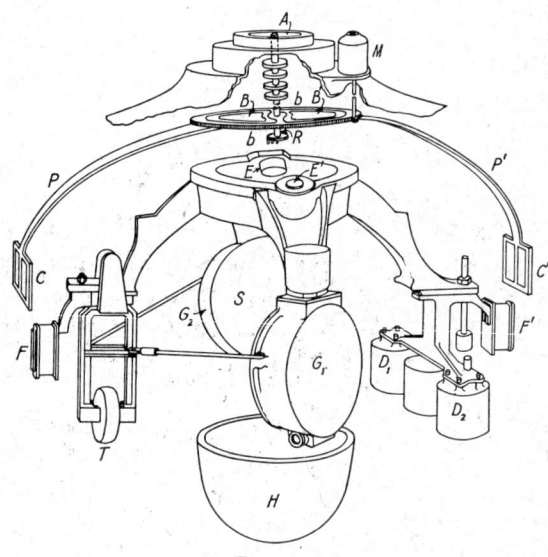

Fig. 192

cavity filled with mercury in a block of insulating material fastened to the supporting globe at its center. The other two pins dip into annular grooves filled with mercury, in the same block.

143. The Follow-Up System. — There is no mechanical connection between the compass card and the sensitive element. When the north-south line of the card becomes out of the meridian plane, it is brought into line with the meridian axis of the sensitive element by a reversible motor M controlled by an alternating current induced in two pairs of coils B, B', C, C', Fig. 192, attached to the phantom. This induced current is due to rapidly changing magnetic fields set up by four alternating current magnets, E, E', F, F', attached to the sensitive element.

The turns of the follow-up coils B and B' are in a horizontal plane above two alternating current magnets E and E' that project upward from the top of the sensitive element. The two coils are wound in opposite directions. The turns of another pair of follow-up coils C, C' are wound in opposite directions in a vertical plane and normally are in front of horizontal alternating current magnets F and F'.

As shown in Fig. 193, a 120-volt, 60-cycle current energizes the primaries of two transformers, the secondaries of which are connected to the four alternating current magnets E, E', F, F' fastened to the sensitive element. Current from the same line is rectified and led to the mid-point of the field coils of the follow-up motor. The amplifying and rectifying circuit consists of three triple-electrode vacuum tubes with the necessary transformer, condenser and resistances. In the diagram is shown an extra set of three tubes to guard against interruption of the operation of the system in case a tube should fail. In the subsequent description, the tube shown at the left end of each set will be called the " input tube " of that set, and the other tubes of each set will be called the first and the second " output " tubes, respectively. A 110-volt direct current energizes the armature of the reversible follow-up motor and also the plate circuit of the input amplifying tubes.

The field coils of the follow-up motor are wound in opposite directions so that, if no current traverses them except the rectified current from the alternating current supply line, the two field pole strengths will be equal and of the same sign. Under this condition there will be zero torque tending to rotate the armature.

The current induced in the follow-up control coils, after being rectified and amplified, is superposed on the rectified current in the follow-up motor field coils from the alternating supply line. The follow-up control coils are connected in the grid circuit of the input tube. The plate circuit includes the primary windings of the interstage transformers. When the ship changes course, an alternating electromotive force is induced in the follow-up control coil connected to the grid circuit. This results in an alternating characteristic being impressed on the direct current flowing in the plate circuit and an alternating electromotive force being induced in the interstage transformer secondaries. These secondary windings are connected with the circuit in such a manner that the grid of one output tube becomes more positive with respect to its filament at a given instant, and the grid of the other tube becomes more negative. Thus the current in one follow-up motor field coil will be strengthened and the current in the other field weakened. Consequently the armatures will rotate and

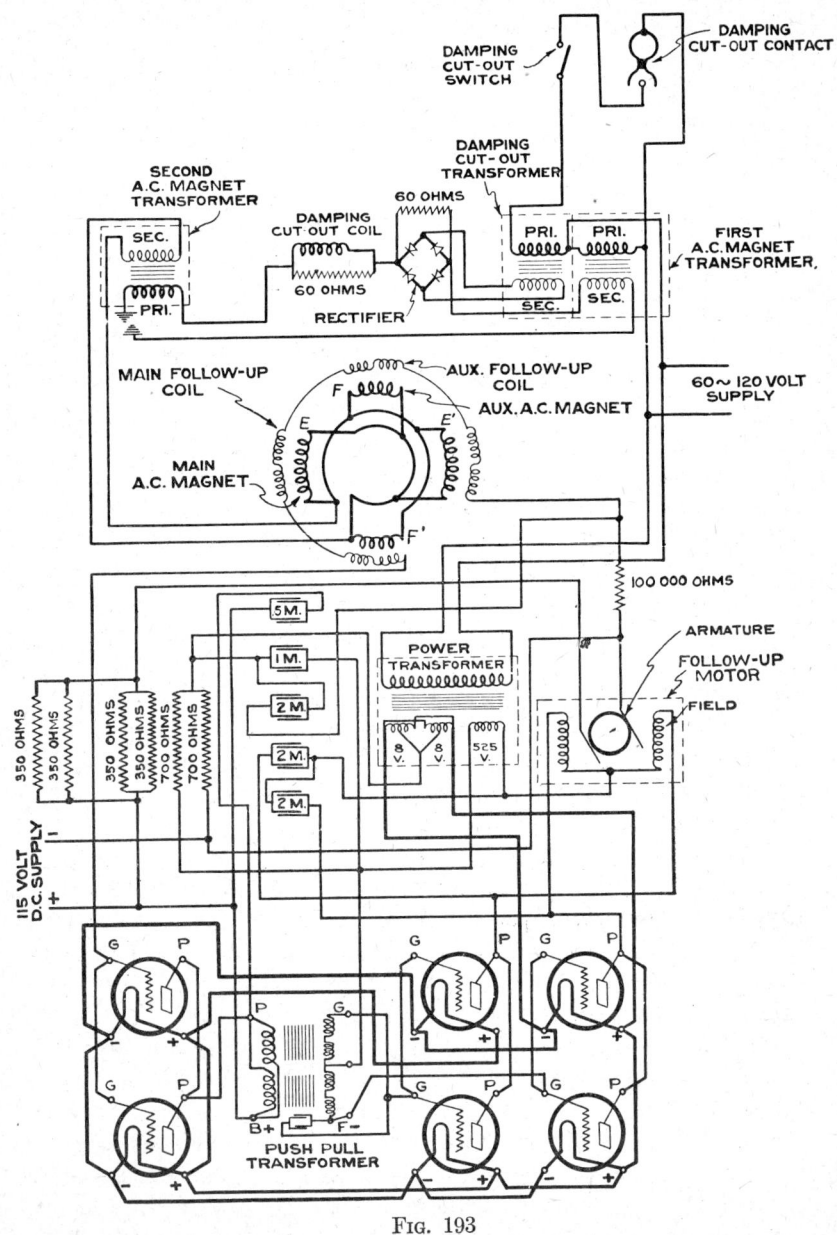

Fig. 193

turn the phantom element till the intersections of the three pairs of follow-up control coils are opposite the corresponding alternating current control magnets. When the ship changes course in the opposite direction, the follow-up motor rotates in the opposite direction thereby again bringing the phantom with the connected compass card into coincidence with the sensitive element.

The pair of horizontal follow-up control coils BB' are of larger diameter and are separated by a wider gap than the two pairs of vertical follow-up control coils C and C'. When the north-south line of the compass card and the connected phantom element are displaced through a considerable angle from the meridian-seeking axis of the sensitive element, the small vertical follow-up control coils will be beyond the range of induction of the follow-up alternating current magnets, but an alternating current will be induced in the large follow-up control coils. This current will cause the follow-up motor to turn the phantom element till the small vertical follow-up control coils are so close to the corresponding alternating current magnets that currents will be induced in their coils also. This action results in the phantom element being turned till the north-south line of the compass card is parallel to the meridian-seeking axis of the sensitive element. The large horizontal follow-up control coils produce a coarse adjustment, and the small vertical follow-up coils furnish the fine adjustment.

Geared to the follow-up system is the transmitter system for the operation of repeater compasses, course recorder and automatic pilot. Also geared to the follow-up system is a device that makes the proper correction for the course and speed error.

144. The Course and Speed Error Corrector. — The gyro-compass on a moving ship is subject to an error which depends upon the latitude and upon the meridian component of the velocity of the ship (Art. 111). The compass is deflected toward the west when the meridian component of the velocity of the ship is directed toward the north, and toward the east when the meridian component is directed toward the south. This error is called the meridian-steaming error, the north-steaming error, and also the course and speed error. The method used in the Arma compass to correct this error consists in setting the phantom element so that the line joining two diametrically opposite follow-up coils makes with the line joining corresponding alternate current control magnets an angle equal to and in the same direction as the deflection of the sensitive element due to the course and the speed of the ship. Then, when the follow-up system turns the phantom, with the connected compass card, till the follow-up coils are in line with the corresponding control magnets, the compass card will show no error due to the course and speed of the ship although the meridian-seeking axis of the sensitive element is deflected from the meridian.

In Figs. 194 and 195 the circle a represents a gear fastened to the axle that carries the compass card; b, a gear fastened to the upper end of the vertical axle of the phantom element; c and d, two gears capable of turning about axles fastened to a plate at-

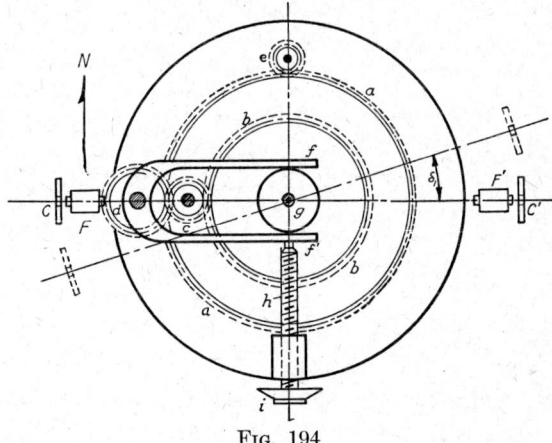

Fig. 194

tached to the compass card; e, a pinion connecting the follow-up motor and the compass card. Fastened to the gear d is a fork ff'. If the prongs of the fork be moved nearer to or farther from e, the attached gear d is turned, and the connected phantom gear b is turned in the same direction. Such a displacement of the prongs of the fork can be produced by moving forward or back a nearly frictionless wheel g on the end of a screw h which is rotatable in a nut fastened to the compass frame.

Suppose that the meridian component of the ship's velocity is directed north. Then, the meridian-seeking axis of the sensitive element will be deflected toward the west through an angle δ_1, depending upon the latitude, speed and course of the ship (124). Values of this angle for various conditions are given in tables. If no correction be made, the follow-up system will rotate the north-south line of the compass card through the same angle in the same direction.

Suppose, now, that the proper angle corresponding to δ_1 be set up on the dial i attached to the screw that moves the wheel g forward or back. For a north meridian component of ship's velocity, the dial-setting will turn the fork and the attached gear d counter-clockwise. The connected phantom gear b will be turned in the same direction, thereby rotating the line connecting the attached follow-up coils CC', Fig. 192, through the angle δ_1 in the

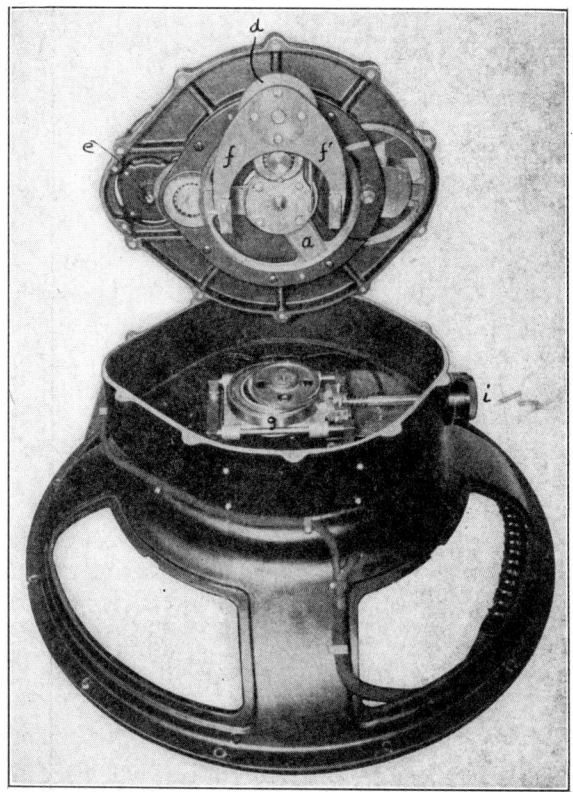

Fig. 195

counter-clockwise direction away from the line connecting the control magnets FF'. Since the line of action of the force with which the screw pushes against the fork passes through the axis of rotation of the compass card, this force is without effect on the indication of the compass card.

Before the dial was set, the line connecting the follow-up coils CC' on the phantom element coincided with the line connecting the control magnets FF' on the sensitive element. Also, both the sensitive element and the compass card were deflected from the meridian through an angle δ_1 in the counter-clockwise direction. The setting of the dial rotated the line connecting the follow-up coils CC' through an angle δ_1 in the counter-clockwise direction away from the line connecting the control magnets FF' without changing the direction of the north-south line of the compass card. The follow-up system immediately started to rotate the pinion e, thereby turning both the compass card and the phantom geared to it through the angle δ_1 in the clockwise direction, that is, till the line connecting the follow-up coils CC' was brought again into coincidence with the line connecting the control magnets FF'. By this rotation of the compass card, the north-south line was brought into the meridian. The meridian-seeking axis of the sensitive element is still at an angle δ_1 to the meridian.

145. Prevention of the Ballistic Deflection Error. — A gyro-compass will be without ballistic deflection error at the equator if the undamped period has a value given by (126). If the compass be moved to any other latitude λ, the compass will continue to be without ballistic deflection error if

$$\frac{K_s w_s}{mx \cos \lambda} [= Rw_e]$$

be maintained constant (Art. 113).

The desired constancy can be maintained by varying the spin velocity w_s as the cosine of the latitude varies. The spin-velocity of the gyros of the Arma compass is changed by adjusting the speed of the motor generator that energizes the stator coils of the two gyro-motors. In Fig. 196, D, E are the terminals of the direct current motor that rotates the three-phase generator XYZ which operates the two gyro-motors G_1 and G_2. The armature A of the direct current motor is connected in series with a variable rheostat R. The rotatable arm of the rheostat is keyed to the shaft of a small reversible two-phase " torque motor " T. The rotor of the torque motor consists of a squirrel cage armature without commutator or slip rings. When the two stationary coils of the motor are traversed by two alternating currents in different phase, a rotating magnetic field is developed which produces a torque on the rotor. Rotation of the rotor is opposed by the torque developed by a spring, one end of which is fastened to the shaft and the other end attached to a fixed support. The rheostat arm turns till these two torques balance one another. The angle of turn is less than one revolution.

The two field coils of the torque motor are connected to a tuned resonant circuit $JKLN$ consisting of a variable inductance M, resistance R_1 and fixed capacitances C_1 and C_2. A choke coil R_2 prevents a too rapid increase of current. The resonant frequency of the circuit is inversely proportional to the square root of the inductance. It is possible to vary the frequency of this circuit within wide limits, and consequently vary the resistance of the rheostat R, the current in the series field coil S of the direct

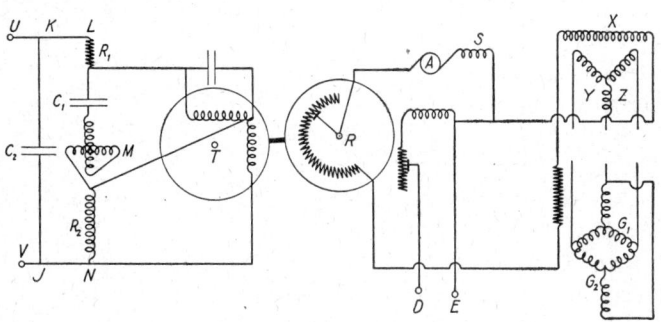

Fig. 196

current motor, the speed of this motor, the electromotive force of the alternator XYZ and therefore the spin-velocity of the two gyros G_1 and G_2.

In setting the dial fastened to the variable inductance M for a particular latitude, the resonant frequency of the tuned circuit is made to correspond to the speed at which the motor generator should run when the compass is at that latitude. If the speed is too low, the reversible torque motor T will turn in the direction to reduce the resistance of the rheostat R. If the speed is too high, the reversible motor will turn in the opposite direction, thereby increasing the rheostat resistance. At the correct motor-generator speed, the potential difference at the terminals of the two field windings of the torque motor are in phase and the torque motor stops.

A tilt of the spin-axle of a gyro-compass is necessary in order that the spin-axle may remain in the meridian when at any latitude either north or south of the equator (Art. 106). In Art. 109 it has been shown that if the spin-velocity of a gyro-compass be considerably altered while the spin-axle is tilted from the horizontal, the spin-axle will deflect in azimuth. The method used in both the Anschütz and in the Arma gyro-compasses for avoiding the ballistic deflection error by changing the spin-velocity of the gyros as the latitude of the ship changes must result in a certain azimuthal deflection. This deflection is reduced to a negligible amount, however, if the proper change in spin-velocity is made corresponding to each small change in latitude. At latitudes in the neighborhood of 40° the azimuthal deflection is about 0.2 degree for changes of spin-velocity corresponding to 5-degree changes in latitude.

146. Avoidance of the Ballistic Damping Error. — When the meridian component of the velocity of the ship changes either in direction or in magnitude, the inertia of the oil in the damping tanks causes oil to move in the direction opposite that of the acceleration of the ship. For example, if a ship while steaming northward either suddenly stops or makes a quick turn, there is an acceleration of the ship toward the south and oil will move from the south damping tank to the northern one. If the oil could flow freely from one tank to the other, the oil would become at the same level in the two tanks when the acceleration ceased. The constricted passage connecting the two tanks retards equalization however, and develops a precession of the sensitive element away from the normal resting position. This results in a ballistic damping or ballistic turning error (Art. 114).

This error is avoided by preventing damping of vibration of the sensitive element during the time that the velocity of the ship is changing. The greatest acceleration produced by any practical change of speed of a ship on a straight course is so small that the ballistic damping error is negligible so long as the course of the ship is unchanged. The acceleration may be so great when a ship steaming at considerable speed makes a sudden change of course that the turning of a ship may produce a large ballistic damping error. All devices for avoiding the ballistic damping error are designed to prevent damping throughout the time that the ship is making a turn.

The Arma damping cut-out device consists of an electric solenoid-operated valve in the oil line between the two damping tanks. The solenoid is energized by a unidirectional intermittent current that is controlled by a contact switch mounted on the shaft of the phantom element. This cut-out contact switch consists of a con-

THE FLORENTIA GYRO-COMPASS 247

tact stud attached to the compass frame and a fork attached to a collar that can turn with a certain amount of friction about the phantom shaft. When the ship starts to turn in either direction, the friction between the cut-out switch and the phantom shaft carries the switch around and brings one prong or the other into contact with the fixed contact stud. This completes the electric circuit through the solenoid and causes the cut-out valve to close. If the ship had turned in the opposite direction, the other prong would have made contact and the valve would be closed as before. The distance between the prongs is such that contact is always made before the ship has turned more than about 10 degrees. After the ship has ceased turning, the yawing of the ship back and forth quickly breaks the switch contact, and normal damping is resumed.

147. Avoidance of the Quadrantal or Rolling Error. — A compass with a single gyro is subject to an error when on a rolling ship on an intercardinal course (Arts. 115, 116). This error is avoided in the Arma, as in the Anschütz gyro-compass, by the use of two gyros with the spin-axles inclined to one another (Art. 140).

§7. *The Florentia Gyro-Compass*

148. Arrangement of the Principal Parts of the Florentia Master Compass. — The sensitive element of this instrument,* made by the Officine Galileo, Florence, Italy, consists of a single meridian-seeking gyro-wheel in a case, G, Fig. 197, hanging from a hollow ring F that floats in an annular trough of mercury T. The gyro has a mass of about 5 kg. and a moment of inertia of about 70,000 gm. cm^2. It rotates at a speed of 20,000 revolutions per minute.

The mercury trough T hangs from a vertical spindle carried by a ball-bearing on the horizontal frame or spider D. The spider is suspended by gimbal rings, R_1R_2, from the top of the binnacle B. The lubber line is engraved on an arm L fastened to the spider. The compass card C is fastened on top of the spindle. The rigidly connected system from the mercury trough to the compass card constitutes the phantom element, P. The phantom element is stabilized by a second gyro SG within a casing directly below the compass card. By means of a reversible motor M, pinioned to a

---

* U. S. Patent. Martienssen, No. 1493213, 1924.

horizontal gear on the phantom element P, the latter is kept in fixed position relative to the meridian-seeking axle of G.

The sensitive element S is kept central with respect to the annular trough by two pins p', p. The upper pin is attached to the phantom and projects into a small ring fastened to the sensitive element. The lower pin is attached to the gyro-casing G and projects into a slot q in the end of an arm fastened to the phantom element. The center of mass of the sensitive element is about

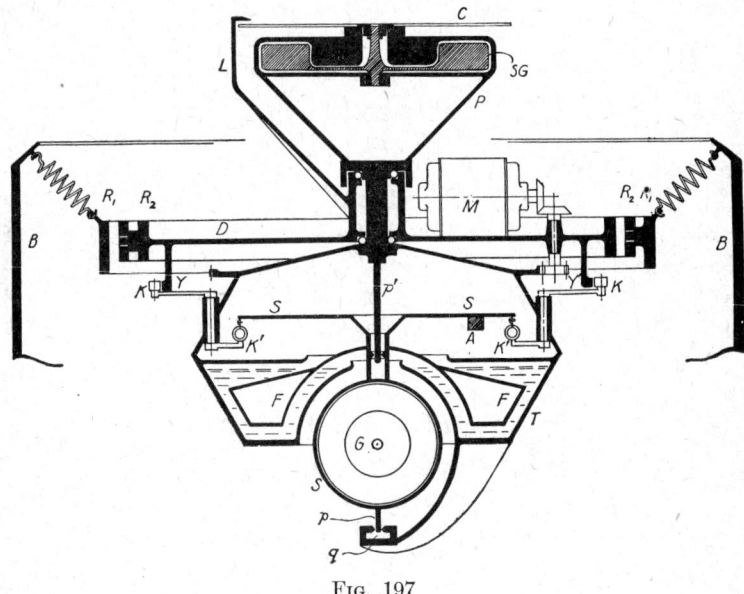

Fig. 197

0.8 cm. below the center of buoyancy. The Florentia gyro-compass is slightly pendulous.

The gyro of the sensitive element is the rotor of a three-phase motor. One of the three phases is led to the motor through the supporting spindle. The other two are led through two flexible platinum strips that extend from the phantom to the gyro-casing.

149. The Follow-Up System. — The upper end of a vertical helical spring is suspended from the upper plate of the sensitive element. On the lower end of the spring is a pair of silver balls, Fig. 198. These balls hang between a pair of vertical metal plates X, X', Fig. 199, mounted on a frame carried by the phantom element.

When the north-south line of the compass card is in the vertical

THE FLORENTIA GYRO-COMPASS 249

plane of the spin-axle of the meridian-seeking gyro, the balls are midway between the two vertical plates. When the phantom turns relative to the spin-axle of the meridian-seeking gyro, one of the balls comes into contact with one of the vertical plates, thereby completing an electric circuit through the follow-up motor

Fig. 198

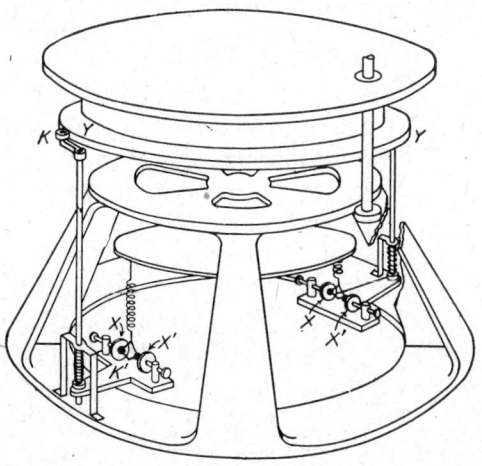

Fig. 199

M mounted on the spider. The motor rotates the phantom till neither ball touches a contact plate, that is, till the north-south line of the compass card is in the vertical plane through the spin-axle of the meridian-seeking gyro. Really, the motor rotates the phantom slightly beyond the equilibrium position, so that the other vertical plate is brought into contact with the hanging balls and the direction of rotation of the motor is reversed. Thus, the compass card is caused to " hunt " back and forth through an amplitude of a degree or two, thereby avoiding any chance for the phantom to stick.

The position of the north-south line of the compass card relative to the lubber line is transmitted to the course recorder, and to repeater compasses, by means of a transmitter commutator as in the case of other gyro-compasses.

150. Damping. — Damping of the oscillations of the spin-axle is effected by a method that causes the follow-up motor to absorb energy from the oscillating element.* The pin p, Fig. 197, which extends downward from the lower side of the gyro-casing projects into a slot q in an arm attached to the phantom element. The east side of the sensitive element is overweighted so that the pin presses against the west jaw of the slot. In Fig. 197, this overweighting is indicated by a mass A on the east side of the sensitive element.

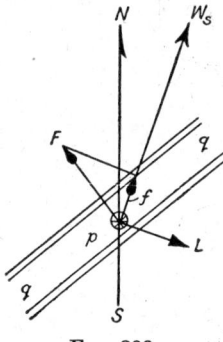

Fig. 200

The spin-axle of a pendulous gyroscope tends to become parallel to the earth's axis, with the direction of spin in the same sense as the direction of rotation of the earth (Art. 106). It follows that when the spin-axle is in equilibrium, the sense of rotation of the gyro is clockwise as seen by an observer looking along the axle from south toward the north.

When the spin-axle of the Florentia gyro-compass oscillates, the phantom follows, in azimuth, the motion of the sensitive element. The north-south line of the compass card and the parallel slot are in the vertical plane of the spin-axle only when the sensitive element is in the resting position. At other times the vertical plane through the slot is slightly behind the vertical plane through the spin-axle. Consider the case when the north-seeking end of the spin-axle is east of and moving toward the meridian plane. In Fig. 200, the spin-velocity is represented by w_s, the cross-section of the vertical pin by p, and the slot in the arm attached to the phantom element by q. Since the east side of the phantom is overbalanced, the east jaw of the slot pushes on the pin with a force represented by F. The component f of this force, parallel to the spin-axle, develops a torque on the sensitive element represented by L. As the spin-axle tends to set itself parallel to the torque-

* Oscar Martienssen, "Eine neue Methode zur Dämpfung der Schwingungen eines Kreiselkompasses," Physik. Zeit. **29**, 295–300 (1928).

axis and with the direction of spin in the direction of the torque, it follows that the oscillation of the spin-axle is opposed. The pin p projecting downward from the gyro-casing is acted upon by forces that oppose both the motion of the spin-axle in azimuth and the motion in elevation. These oppositions to the motion of the spin-axle produce rapid damping of the vibration of the sensitive element.

If the sensitive element were overbalanced on the west side, the amplitude of successive swings of the spin-axle of the meridian-seeking gyro would become greater and greater.

151. The Latitude and Meridian-Steaming Error Corrector. — Since the method employed to damp the vibration of the sensitive element of the Florentia gyro-compass opposes tilting of the axle of the meridian-seeking gyro, this compass is subject to the latitude error (Art. 110). In common with all gyro-compasses it is subject to the meridian-steaming error (Art. 111). These two errors are corrected by a device that turns the phantom, relative to the spin-axle of the sensitive element, through such an angle that the north-south line of the compass card is free of both errors although the spin-axle of the sensitive element is not in the meridian plane.

Fig. 201.

The pair of silver balls, Fig. 198, suspended from the top of the sensitive element, hang between two vertical plates, X, X', on one end of a crank, KK', Figs. 199 and 201, mounted on a frame attached to the phantom. A helical spring causes the upper end of the crank to press against the edge of a ring, Y, Figs. 197 and 199, fastened flat against the under side of the spider. The position of the ring can be adjusted so as to be eccentric to the vertical axis of the sensitive element and phantom. This eccentric ring constitutes a cosine cam similar to the one used on the Sperry Mark VI and Mark VIII gyro-compasses (Art. 123). The degree of eccentricity is controlled by the position of an arm on the upper face of the spider. This correction arm is graduated for latitude and is capable of being moved over a flat scale carrying speed curves.

When the correction arm is set for a given latitude, speed, and course, the cam determines the proper position of the crank, and consequently the position of the contact plates, relative to the north-south axis of the compass card. The follow-up motor, in circuit with the contact plates and the pair of balls between them, now maintains such an angle between the north-south axis of the compass card and the spin-axle of the sensitive element that the latitude and the meridian-steaming errors of the sensitive element do not appear in the compass card indication. The probability of any failure to operate is reduced by the addition of a duplicate corrector system consisting of a second cosine cam Y' together with an additional pair of contact plates and pair of silver balls placed diametrically opposite the ones just described.

152. Avoidance of the Ballistic Deflection Error. — In the Florentia, as well as in all other gyro-compasses, the deflection of the spin-axle of the sensitive element from the meridian, that would be produced when there is a change in either the speed or course of the ship, is avoided by designing the instrument so that the period of the undamped azimuthal vibration back and forth is about 84 minutes when at the equator (Art. 113). At other latitudes the period should be less. The period of the Florentia gyro-compass is fixed by the makers so as to reduce to a negligible value the ballistic deflection error in the region in which most sailings are to be made. There is no device on the gyro-compass of 1924 by which the navigator can alter the period, nor any device for avoiding the ballistic damping error (Art. 114).

153. Avoidance of the Error Due to Rolling and Pitching of the Ship When on Intercardinal Courses. — Deflection of the spin-axle of the sensitive element from the meridian produced by rolling and pitching of the ship is much reduced by keeping the entire suspended system always nearly vertical, however the ship may roll or pitch (Arts. 115 and 117). This result is accomplished by making the phantom into a gyro-pendulum of long period.* The period of the conical oscillation of the phantom element about a vertical axis is increased to about 40 seconds by means of a gyro, SG, Fig. 197, having a mass of about 10 kg. and a moment of inertia of about 405,000 gm. cm.^{2} This stabilizing gyro not only increases the period of vibration of the phantom but also greatly reduces the amplitude of vibration. A top view of the entire instrument is given in Fig. 202.

The stabilizing gyro opposes any torque that tends to rotate the phantom about any axis inclined to its spin-axle. Consider

* U. S. Patent. Martienssen, No. 1493214, 1924.

the effect of a torque that tends to rotate the phantom element about an axis perpendicular to the plane of the diagram, Fig. 197. Such a torque causes the spin-axle of the stabilizing gyro to precess slowly about a horizontal axis in the plane of the diagram. The annular mercury trough is thereby slightly tilted but the mercury

FIG. 202

surface remains practically horizontal. Consequently, the direction of the meridian-seeking gyro axle is unaffected by the precession or by the torque that produced it.

QUESTIONS

1. Would a gyroscope which is entirely free from friction about its separate axes align its spin-axis with the rotational axis of the earth when the gyro-axle is given an initial easterly or westerly displacement? Explain.

2. What are the controlling forces which convert a gyroscope into a meridian-seeking device?

3. Show why the spin-axle of a pendulous gyro-compass tends to set itself in the meridian plane and with the direction of spin in the same direction that the earth rotates, that is, clockwise as seen from the south.

4. Show why the spin-axle of a liquid-controlled non-pendulous gyro-compass tends to set itself in the meridian plane and with the direction of spin in the direction opposite the rotation of the earth, that is, counter-clockwise as seen from the south.

5. Show that each end of the spin-axle of an undamped gyro-compass moves in an elliptical path.

6. Describe three methods employed to damp the vibration of gyro-compasses back and forth across the meridian plane.

7. Show that the damping device used on the Anschütz gyro-compass is applicable only to a compass that spins in the same direction as the rotation of the earth.

8. Why are the spinning wheels of gyro-compasses always of great angular momentum?

8. Deduce an expression for the magnitude of the directive force acting upon a gyro-compass. Where, upon the earth's surface, does the gyro-compass possess its maximum and where its minimum directive force?

10. Suppose that, in assembling a gyro-compass, the gyro-wheel were mounted slightly too far toward the south. What would be the effect upon the direction of the settling position of the spin-axle of the gyro?

11. Show that if the spin-velocity of a Sperry gyro-compass diminishes considerably from the correct value, the compass will deflect westward of the meridian plane when the compass is in northern latitudes.

12. In going from the equator to latitude 60°, the spin-velocity of both the Anschütz and the Arma gyro-compass is reduced 50 per cent. Why? State the reason why this change is not accompanied by a deflection of the spin-axle out of the meridian plane.

13. Describe the cause of the "latitude error." Give the names of the gyro-compasses in which it exists and show how it is compensated.

14. Describe the cause of the "meridian-steaming" or "north-steaming" error and show how it is compensated.

15. Describe the cause of the "ballistic deflection" error and show that this error will not occur if the period of vibration of the spin-axle has the proper value.

16. Find the period of a gyro-compass that will have zero ballistic deflection error when the compass is at a given latitude.

17. Show what is meant by the "ballistic damping error" and state how it can be prevented.

18. Describe the cause of the "quadrantal" or "rolling" error and give two methods by which it is suppressed.

19. The mercury ballistic of the Sperry gyro-compass is supported by a "phantom element." Why could it not be attached directly to the gyro-casing?

20. The mercury ballistic of the Sperry gyro-compass is loosely connected to the gyro-casing by a pin that is a little to the east of the vertical line through the center of the gyro. Show what would be the effect upon the compass if this eccentric connection were offset to the west of the vertical axis rather than to the east.

21. Describe the methods used in three different makes of gyro-compass for causing the position in azimuth of the compass card of the master compass to remain in fixed relation with respect to the direction of the spin-axle of the gyro.

22. Name a natural error to which gyro-compasses are subject, that is avoided in each of the following types: (*a*) pendulous sensitive element, (*b*) non-pendulous element provided with a mercury ballistic, (*c*) **two-gyro** sensitive element. Explain how each type avoids the error named.

CHAPTER VI

GYROSCOPIC STABILIZATION

§1. *General Principles*

154. Static and Kinetic Stability. — If a body, after suffering a slight angular displacement, recovers its former position, the body is said to be *stable*. If, after suffering a slight angular displacement, the body departs further and further from its former position, the body is said to be *unstable*. If, however, after suffering a slight angular displacement, there is no tendency of the body either to recover its former condition or to depart further from it, the body is said to be in *neutral* or *indifferent stability*. A body that is stable is also said to have positive stability and one that is unstable is said to have negative stability. The degree of stability of a body is measured by the amount of work necessary to effect a permanent change in the position of the body.

When a whipping-top is standing on its flat end it is statically stable; when standing on its peg and not spinning, it is unstable; when lying on its side, it is neutral. If, however, the top be given an angular displacement when spinning with constant speed, the top will not tumble over as it would if not spinning but will oscillate and eventually regain very nearly the condition it had before it was disturbed. A top, when spinning, is said to be kinetically or dynamically stable. Since, in this case, the top is acted upon by an unbalanced torque, the top is not in equilibrium.

A steady motion is said to be *kinetically* or *dynamically stable* if, when slightly disturbed, it oscillates in such a manner that the vibratory deviation from steady motion approaches steady motion as a limit when the disturbance approaches zero.

155. The Stability of a System Consisting of a Body Capable of Oscillation and an Attached Precessing Gyro-Wheel. — Represent the mass of the system by m, the radius of gyration with respect to the axis of oscillation by k, and the distance from the center of mass to the point of support by H. Represent by θ the angle at any instant between any line fixed in the system and the position of this line when the system is in its equilibrium position.

(a) *A Statically Stable Gyro on a Statically Stable Body Capable of Oscillation under the Influence of Gravity.* — When the amplitude of oscillation θ is small and the gyro-wheel is not spinning, the equation of motion of the system of mass m is

$$mk^2 \frac{d^2\theta}{dt^2} = -mgH\theta \qquad (131)$$

where t is the time occupied in moving through the angle θ.

If the gyro-wheel is spinning, there is an additional torque acting on the system (Art. 69) in the same direction as that due to gravity. When the spin-axis is perpendicular to the axis of oscillation, the equation of motion of the system is (90) and (131),

$$L_c' = -mgH\theta - \frac{K_s^2 w_s^2 \theta}{K_c} \qquad (132)$$

where K_c represents the moment of inertia of the gyroscope with respect to the axis about which the system oscillates.

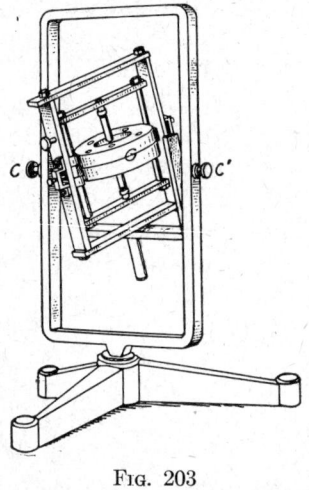

Fig. 203

Since no change of the magnitude of θ produces a reversal of the direction of the resultant torque acting on the system, it follows that the system is not only statically stable but also always dynamically stable.

(b) *A Statically Unstable Gyro on a Statically Unstable Oscillating Body. Experiment.* — Figure 203 represents a system including a gyro-wheel G that can either hang pendulously and oscillate about a rod cc' or can be turned into the position shown in the figure with the center of gravity above the axis of support. When in the position shown in the figure, if the system be slightly displaced about the axis cc' while the wheel is not spinning, the system will continue to rotate for 180 degrees till the center of gravity is below the line cc'. The system is sometimes called a stilt top or inverted gyroscopic pendulum. It is statically unstable when in the position shown.

With the gyroscope anti-pendulous and the entire system unstable, set the gyro-wheel spinning. Now push on the gyro-frame, thereby producing a small displacement about the axis cc'. The system does not flop over. It is dynamically stable.

In the present case, the torque due to gravity is in the same direction as the gyroscopic torque which was developed by the push and which produced an angular acceleration about the axis cc'. The torque acting on the system is, (Art. 69):

$$L_c'' = mgH\theta - \frac{K_s^2 w_s^2 \theta}{K_c}$$

The direction of the resultant torque depends upon the value of θ. This system is dynamically stable when

$$\frac{K_s^2 w_s^2 \theta}{K_c} > mgH\theta$$

Multiplying each side of this inequality by $\frac{K_c}{2K_s}$, we see that the condition of dynamical stability assumes the form:

$$\frac{K_s w_s^2}{2} > \frac{K_c}{K_s}\left(\frac{mgH}{2}\right)$$

The left-hand side of this inequality represents the kinetic energy of rotation of the gyro-wheel. The quantity within the parenthesis represents one-half of the change in the potential energy of the system in falling through the distance H. Thus, an oscillating system that is statically unstable can be rendered dynamically stable by means of a statically unstable gyroscope capable of precessing about an axis perpendicular to the axis of oscillation of the system, if the kinetic energy of rotation of the gyro-wheel is greater than K_c/K_s times one-half of the potential energy lost by the system in rotating about the axis of oscillation from a vertical position to a horizontal position.

This principle is the basis of all methods proposed for stabilizing monorail cars.

156. Some Laws of Dynamic Stability. — There are several laws of dynamic stability to which we shall need to refer in subsequent considerations. The following can be proved by analytic methods and can be illustrated readily by the apparatus shown in Fig. 203.

(a) If the inner frame and spinning wheel of a gyroscope be statically stable with respect to a horizontal axis perpendicular to the spin-axle, and if at the same time the second frame be also statically stable with respect to a horizontal axis perpendicular to the first, then the system will be dynamically stable and the two axes will precess in the direction opposite the spin (Art. 155, a).

(b) If the inner frame and spinning wheel of a gyroscope be statically unstable with respect to a horizontal axis perpendicular to the spin-axle, and if at the same time the second frame be also statically unstable with respect to a horizontal axis perpendicular to the first, then the system may be dynamically stable, and when it is dynamically stable the two axes will precess in the direction of the spin (Art. 155, b).

(c) Precession of a statically stable gyroscopic system is suppressed by retarding the precession till the system is in its equilibrium position.

(d) Precession of a statically unstable gyroscopic system is suppressed by accelerating the precession till the system is in its equilibrium position.

(e) Hurrying the precession of a top causes the top to rise against the torque due to gravity. Hurrying the precession of a statically unstable gyroscope causes the spin-axle to rotate about the axis of torque in the direction opposite that of the torque (Art. 40).

(f) Retarding the precession of a top causes the top to fall. Retarding the precession of a statically unstable gyroscope causes the spin-axle to rotate about the axis of torque in the direction of the torque (Art. 40).

(g) A gyro-wheel incapable of precessing produces no effect on the stability of a body to which it may be attached (Art. 34).

(h) A body that is statically stable and subject to periodic oscillations can be rendered dynamically stable by retarding the precessional velocity of the vertical gyro-axle of an attached gyroscope that is either statically stable or neutral (Art. 88).*

(i) A body that is statically unstable and subject to oscillations can be rendered dynamically stable by accelerating the precessional velocity of the vertical gyro-axle of an attached unstable gyroscope.*

(j) A body consisting of two gyroscopically coupled systems capable of oscillating about horizontal axes cannot be dynamically stable unless both are either statically stable or statically unstable.†

(k) The overloading of one side of a system that is statically unstable, but that has been rendered dynamically stable by means of a spinning gyroscope, causes the overloaded side to rise and the center of gravity of the system to oscillate across a vertical line through the point of support of the system.

* Bogaert, L'Effet gyrostatique et ses applications (1912), Art. 68.
† Deimel, Mechanics of the Gyroscope (1929), Art. 101.

(*l*) When a statically unstable body rendered dynamically stable by a statically unstable gyroscope with vertical spin-axle is moved around a curve in the direction of the spin of the gyro-wheel, the system becomes more stable; when revolved in the direction opposite the spin, the system becomes less stable (Art. 159).

(*m*) When a statically unstable body to which is attached a gyroscope of neutral stability and with vertical spin-axle goes around a curve in the direction of the spin of the gyro-wheel, the entire system becomes both statically and dynamically unstable. When the system goes around the curve in the direction opposite the spin, the system oscillates about an equilibrium position.

(*n*) When a statically stable body to which is attached a gyroscope with vertical spin-axle and positive static stability is rotated in the direction of the gyro-wheel, the stability of the system is increased; whereas if the rotation be in the direction opposite the spin, then the stability of the system is decreased.

§2. *Gyroscopically Stabilized Monorail Cars*

157. The Economy of Monorail Cars. — Cars that would safely run at high speed on a single rail on the ground would effect considerable economies over birail cars, not only in original cost of installation but also in maintenance of way and in operation. The ability of a birail locomotive to go up a grade is limited by the coefficient of static friction between the edge of the wheels and the top of the rails. Grades are seldom over two per cent. Five per cent is considered very high. Wheels with double flanges, slightly tapered, would give a greater frictional force but could not be used on a curve with birail cars. They could, however, be used on even a sharp curve with monorail cars, thereby enabling steeper grades to be traversed by monorail cars than by birail cars. A monorailway could be built in a terrain requiring curves and grades too sharp for an ordinary birail railway. There would be many places where a monorailway would be used if we were certain as to the dynamic stability of the statically unstable cars. A considerable amount of thought has been devoted to the dynamic stabilization of monorail cars but at the present time there is no commercial line in operation.

158. The Principles upon Which Depend the Dynamic Stabilization of Monorail Cars. — The dynamic stabilization of a monorail car requires that a torque acting upon the car from outside shall be neutralized by a torque produced within. In all methods thus far used for causing a monorail car to erect itself after becoming tilted from the vertical, the tilting causes the spin-axle of a gyroscope to precess and the velocity of precession is caused to be accelerated by some torque. The differences in the various schemes have been in the number of gyroscopes employed, the direction of the spin-axles relative to the rail and in the method used to produce the acceleration of the spin-axles.

Consider a monorail car in which is mounted a single gyroscope with vertical spin-axle capable of precessing about an axis transverse to the car, Fig. 204. Suppose that the gyro and casing are mounted so as to constitute a statically unstable system. Just as soon as the car tilts from the vertical, the spin-axle

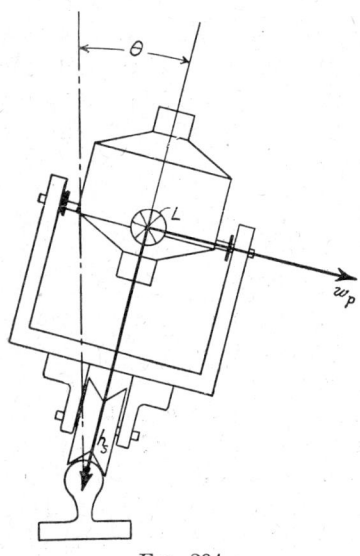
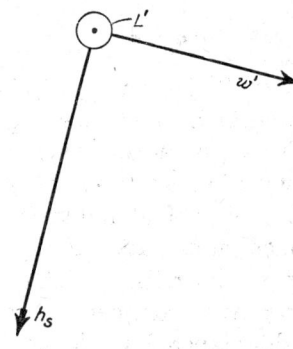

Fig. 204   Fig. 205

begins to precess. If the spin velocity is in the direction indicated by h_s, the direction of the precession is as indicated by w_p. Just as soon as the gyro-axle is deflected from the vertical, the gyro and casing begin to fall in the direction of the gyroscopic precession, thereby increasing the velocity about the precession axis in a given time by an amount w', Fig. 205. A righting torque L' now acts on the car (Art. 36). If this righting torque is large enough, it brings the car back toward the upright position in spite of the gravitational torque that tends to tip it over further. But if the righting torque is large enough to bring the car to the upright position it is likely to carry it beyond that position.

The action is similar if the velocity of the spin-axle about the precession axis is made greater than the precession velocity due to the gravitational torque. The angular velocity of the spin-axle may be increased by a motor or other outside agent. There is now an unneutralized righting torque, L''. For maximum effect, the acceleration of the precessional velocity produced by the motor should stop when the free precession stops. The direction of the acceleration should reverse when the direction of precession reverses. The righting torque L'' should comprise two components, one that is proportional to the instantaneous precessional velocity, and one that is proportional to the angle of precession at that instant.

159. The Effect of a Change in Linear Velocity on the Stability of a Monorail Car that Carries a Single Statically Unstable Gyroscope with Vertical Spin-Axle. — First, consider the effect of a change in the magnitude of the velocity of the car. If the speed of the car be increased when moving from left to right on a straight rail, Fig. 206, the center of gravity of the gyro-system will hang backward causing the upper end of the gyro-axle to tilt backward. The tilting is in the direction indicated by w. Suppose that the gyro spins in the direction indicated by h_s. Accompanying

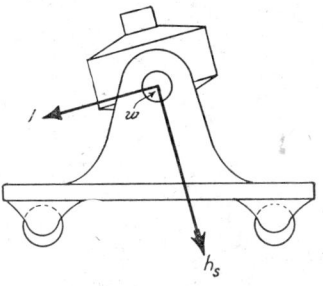

Fig. 206

the tilting backward of the spin-axle is a torque acting on the gyro in the direction indicated by L (Art. 36). This torque tilts the gyro and the attached car to one side of the vertical through the rail. If the linear velocity of the car be decreased, the spin-axle will tilt forward and the car will tilt in the direction opposite that when the velocity of the car is increased. When the linear velocity of the car becomes constant, the car will oscillate from side to side about its position of equilibrium.

Now, consider the effect upon the stability of the car produced by motion around a curve. Figure 207 represents a monorail car, going away from the reader and carrying a statically unstable gyroscope. If the gyro were not spinning and the car started to go around a curve, the car would tilt away from the center of the curve, thereby exerting a gravitational torque L on the gyro. If the gyro is spinning in the direction indicated by the arrow h_s, this torque causes the spin-axle to precess in the direction indicated by the arrow w_p. Suppose that the angular velocity of the spin-axle about the precession axis is accelerated on account of the static unstability of the gyroscope and that the increase in angular velocity of the gyro about the precession axis due to this cause is w', then there is a torque L', Fig. 208, tending to tilt the car toward the center of the curve (Art. 36). This torque is due to the displacement of the center of gravity of the car and that of the gyroscope from the vertical through the rail. If the gyro has a suffi-

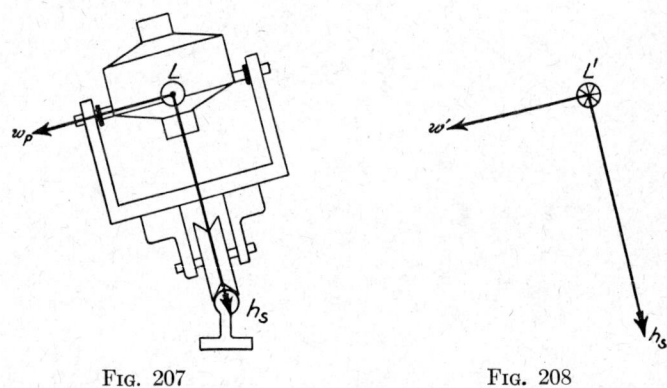

Fig. 207　　　　　　　　　Fig. 208

ciently great angular momentum h_s, the car will lean beyond the upright position toward the center of the curve.

The degree of kinetic stability of the car effected by the gyroscope depends upon the direction and magnitude of the velocity of the car around the curve. Figures 209 and 210 represent the car going away from the reader and making a right turn around a curve with center at the right-hand side of the diagram. Figure 209 represents the car tilting away from the center of the curve and Fig. 210 represents the car tilting toward the center of the curve. The angular velocity w of the car about the curve is in the same sense as the vertical component of the spin-velocity of the gyro. The component of the angular momentum of the gyro in the direction perpendicular to the axis of turning is h_s'. When this component is turned with angular velocity w, a torque L acts on the gyro-wheel in the direction tending to rotate the spin-axle toward a vertical position (Art. 36). As the gyro cannot rotate about the axis of L independently of the car, this torque acts on the car and opposes the tilting of the car from the vertical.

Figures 211 and 212 represent the car going away from the reader and making a left turn around a curve with center at the left-hand side of the diagram. The angular velocity w of the car about the curve is in the opposite sense to the vertical component of the spin-velocity of the gyro. Following the method of

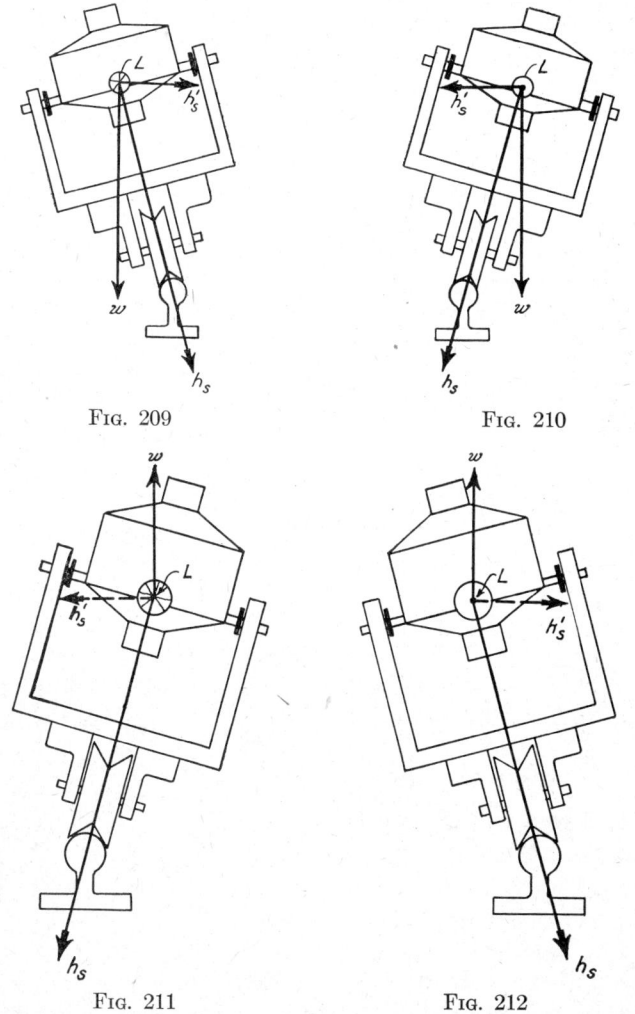

Fig. 209 Fig. 210

Fig. 211 Fig. 212

the preceding paragraph, we arrive at the conclusion that in this case a torque acts on the car tending to increase the tilt from the vertical.

Thus it appears that a monorail car stabilized by a gyro-wheel with vertical axle tilts whenever the velocity of the car is changed either in magnitude or direction. The tilt is greater when the car moves around a curve in the direction opposite that of the spin of the gyro-wheel than when it moves around a similar curve in the same direction as the spin of the gyro-wheel.

160. The Effect of a Change in Linear Velocity on the Stability of a Monorail Car that Carries a Single Gyroscope with Horizontal Spin-Axle Transverse to the Car. — When the car goes

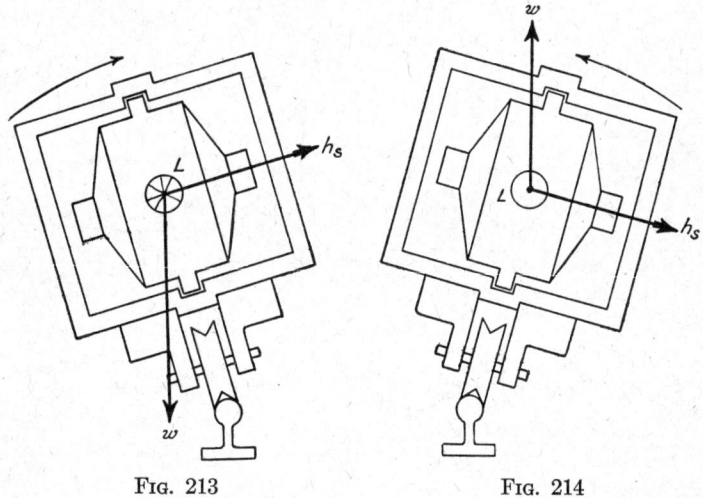

Fig. 213   Fig. 214

around a curve and the gyro is not spinning, the car tilts away from the center of the curve. When the car is making a right turn, if the direction of spin of the gyro is as indicated by h_s, Fig. 213, a torque acts on the gyro in the direction indicated by L (Art. 41). Since the gyro cannot turn, relative to the car, about the axis of this torque, the car is acted upon by a torque which tends to diminish the tilt away from the center of the curve. If the direction of spin of the gyro were reversed, the car would be acted upon by a torque tending to increase the tilt away from the center of the curve.

When the car is making a left turn, if the direction of spin of the gyro is as indicated by h_s, Fig. 214, a torque acts on the car which tends to diminish the tilt away from the center of the curve. If the direction of spin were reversed the car would be acted upon by a torque tending to increase the tilt away from the center of the curve.

161. Methods for Increasing the Kinetic Stability of a Monorail Car while the Car is Going around a Curve. — A monorail car can be maintained kinetically stable if at every instant the precessional velocity of the statically unstable gyro is of the proper value and in the proper direction. In order that a curve may be traversed safely by a monorail car, the precessional velocity of the spin-axle must have a value that is different from that required when the car is moving along a straight level track. The precessional velocity can be changed by applying a torque about the axis of precession. A variable torque may be applied by an outside motor as in the case of the active type of ship anti-roll device (Art. 93). A variable torque may be produced also by varying the degree of statical instability of the gyro system.

Statical instability of the gyro system can be produced either by having the point of support below the center of gravity or by applying springs that tend to pull over the system. The free precession of the gyro-axle and the tilting of the car are slower when the angular momentum of the gyro is increased, and quicker when the degree of static instability of the gyro system is increased.

Mr. Peter P. Schilovsky, who has devoted much attention to monorail cars, produces the required degree of kinetic stability of the car while it is making a turn by altering the degree of static instability of the gyroscope according to the angular velocity of tilting of the car.*

Another method for avoiding any change in the degree of kinetic instability of a monorail car when changing the direction of motion is to employ two gyros spinning in opposite directions about axes which normally are perpendicular to the track. The normal direction of the spin-axles may be either vertical or horizontal. A view of the first case as seen when looking down on a horizontal plane is represented in Fig. 215. The two gyro-casings, with the gyro spin-axles vertical, are represented by G and G'. Suppose that, owing to motion around a curve, wind pressure on the car, ununiform loading, or any other cause, the side A of the car is moved downward. Torques L and L' act on the gyros G and G', respectively. When spinning in the directions indicated by the symbols h_s and h_s', the gyros will precess in opposite directions as indicated by w_p and w_p'. The two gyro-casings are coupled together by two gear segments g and g' so that the precessional

* Schilovsky, The Gyroscope: Its Practical Construction and Applications, p. 224, Spon and Chamberlain, London and New York.

speeds at any instant are equal in magnitude. If, now, these precessional speeds be accelerated to the same degree, equal torques will be developed on the two gyros and on the attached car in directions opposite those indicated by L and L' (Art. 36). If the angular momenta of the gyros and the accelerations of the precessional velocities be sufficiently great, the car will be righted.

The case in which the spin-axles are normally horizontal is

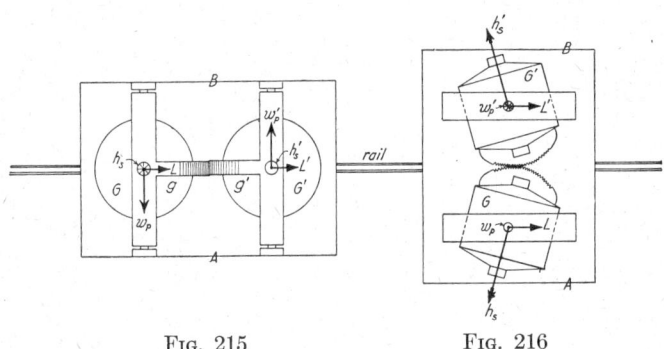

Fig. 215   Fig. 216

represented in Fig. 216. As in the preceding case, the two gyro-axles are coupled together by two gear segments so that the precessional velocities are equal in magnitude and opposite in direction. Suppose that owing to motion around a curve, or any other cause, the side A of the car is moved downward. Then the two gyro-axles will precess in the directions indicated by w_p and w_p', respectively. If, now, these precessional velocities be accelerated to the same degree, equal torques will be developed on the two gyros, and on the attached car, in the directions opposite those indicated by L and L', Fig. 216. If the constants of the gyros are of the proper value, the car will be righted.*

162. The Schilovsky Monogyro Monorail Car of 1915. — Consider a monorail car on which is mounted a single statically unstable gyro spinning about an axis that normally is vertical and capable of precession about a horizontal axis transverse to the rail. When the car tilts to one side, the spin-axle of the gyro precesses either forward or backward. Suppose that a motor accelerates the precession. There is now a torque which opposes the present tilting of the car and tends to make the car tilt in the opposite direction (Art. 158). This overbalancing torque sets up a counter-

* Cousins, " The Stability of Gyroscopic Single Track Vehicles," Engineering (1913), pp. 678, 711 and 781, is a noteworthy analytical treatment.

active force which tends to rotate the spin-axle in the direction opposite to the precession and beyond the vertical position. When the car leans in the direction opposite the original tilt, the direction of precession is reversed. Suppose that now the speed of precession in this direction is accelerated by torques which the motor applies to the gyro. The above series of operations is repeated and the car oscillates from one side of the vertical to the other. If the impulses on the gyro due to the motor are applied at the proper times, are in the proper directions and are of the proper strengths, the motor will absorb energy from the oscillating car,

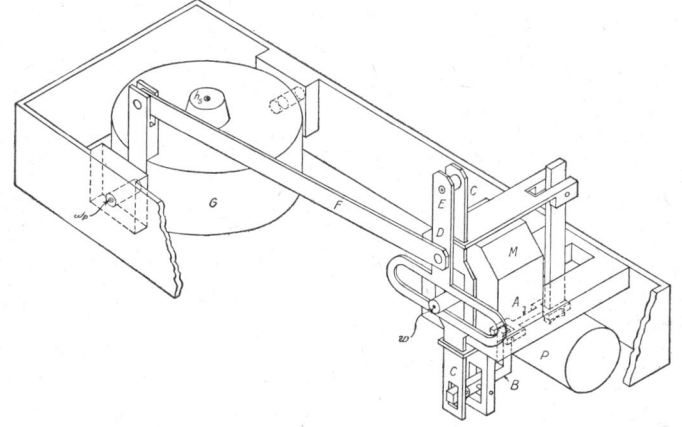

Fig. 217

damp the amplitude of the oscillations and cause the car to become kinetically stable.

The device for stabilizing monorail cars proposed in 1915 by Schilovsky* is represented diagrammatically in Fig. 217. In this much simplified diagram, the various parts of the device are not drawn to scale, the gyro being relatively much larger than here indicated. The gyro G spins about an axis that normally is vertical. It is mounted so as to be statically unstable and capable of precessing about a horizontal axis transverse to the rail. The shaft of the precession motor M rotates continuously with constant speed.

Suppose that the precession motor shaft rotates in the clockwise direction and that the direction of spin of the gyro is as indicated by the symbol h_s. When the car tilts toward the reader,

* U. S. Patent. Schilovsky, No. 1137234, 1915.

the spin-axle precesses in the counter-clockwise direction as indicated by the symbol w_p. At the same time a heavy plumb-bob P pushes toward the reader the link A and the vertical arm of the bell crank B, thereby pushing downward the yoke C and the attached slotted segment D which is fastened to C by a pivot E. This action causes gear teeth on the upper edge of the segmental slot to engage with a gear on the shaft of the precession motor, thereby causing a thrust on the link F and an acceleration of the precession of the spin-axle. The accelerated precession develops a torque that tilts the car to the other side of the vertical. The precession of the spin-axle now reverses in direction, the plumb-bob pulls on the bell crank in the opposite direction and the slotted segment disengages from the motor shaft. Immediately afterward the teeth on the lower edge of the segmental slot engage the gear on the motor shaft, thereby accelerating the present precession of the gyro spin-axle and developing a torque that again causes the car to tilt upward and beyond the vertical. Thus the car oscillates back and forth through the vertical, losing energy of vibration to the precession motor.

With this device, if the constants of the apparatus are of the proper values, the tilt produced when going around a curve is overcome to the same degree and in the same manner as a tilt due to wind pressure or any other cause when the car is moving along a straight level track.

163. The Brennan Duogyro Monorail Car. — The first monorail car stabilized by two gyros, spinning in opposite directions about axes that normally are perpendicular to the track, was designed and built by Louis Brennan. The stabilizing system of Brennan's monorail car of 1905 consists of two gyros G and G', Fig. 218, spinning about axes that normally are horizontal and transverse to the car.* Each gyro is mounted in a casing capable of precessing about a vertical axis. The two casings are coupled together by two segments of gears g and g' so that at any instant the precessional velocities of the two gyros will be equal and in opposite directions. The coupled casings are mounted in a frame capable of rotation through a small angle about an axis through the center of gravity C of the gyro system and parallel to the track. The spin-axle of each gyro carries a rigidly attached roller a, a', and a loose roller b, b', respectively. At a short distance below each of the

* U. S. Patent. Brennan, No. 796893, 1905. Reveille, Dynamique des Solides (1923), pp. 398–406.

keyed rollers a and a', is a short shelf d and d', respectively. Under each of the loose rollers b, b', is a similar shelf e and e', respectively. A plan view of these shelves is shown in the lower part of the diagram. When the car floor is horizontal, the spin-axles are perpendicular to the rail, and each of the keyed rollers a and a' is almost in contact with the corresponding shelf d and d', respectively, at a point near one end of the shelf. The other shelves do not extend

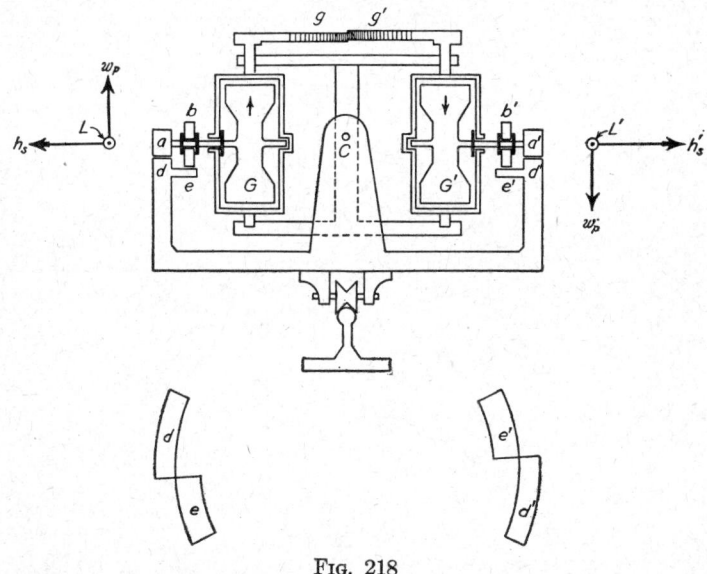

Fig. 218

far enough to be under the loose rollers when the spin-axles are perpendicular to the rail.

Consider a monorail car with the gyros spinning in the directions indicated by h_s and h_s', respectively, Fig. 218. Suppose that for any reason the car tilts to the left. This tilt produces a torque on the spin-axles of the two gyros in the directions indicated by L and L', respectively. The spin-axles precess in the directions indicated by w_p and w_p', respectively. The tilting of the car to the left brings the shelf d' into contact with the keyed roller a'. Because of the friction between this roller and shelf, the roller advances along the shelf toward the reader, thereby accelerating the precession w_p' of the gyro G' and also accelerating the precession w_p of the coupled gyro G. This acceleration is accompanied by a torque on the two gyros in the directions opposite the torques L and L' that produced the original precession (Art. 40). This torque is transferred to the attached frame and tends to right the car. The magnitude of this righting torque depends upon the angular momentum of the gyros and the acceleration of their precessional velocity.*

While the keyed roller a' is moving in contact with the shelf d' toward the reader, the other keyed roller a is moving off the shelf d toward the reader. The loose roller now is above the shelf e but not touching it. The righting torque tilts the car to the right beyond the vertical position, thereby causing the shelf e to press upward against the loose roller b. The torque thereby produced causes the roller a to precess away from the reader and to bring it onto the shelf d. At the same time, the gearing moves a' and b' also away from the reader. During this motion neither a' nor b' touches a shelf. This back-and-forth motion of the car is repeated through smaller and smaller amplitudes of oscillation till the car stands upright or till the car is acted upon by another disturbing torque.

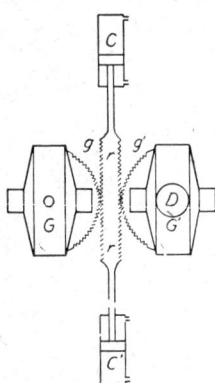

Fig. 219

Brennan's monorail car of 1916 is provided with a more effective device to control the precessional velocity of the gyros. Since the velocity of precession is proportional to the torque tending to tilt the car, the righting torque should be proportional to the precessional velocity. Also, the righting torque should be proportional to the length of time that the disturbing torque acts. In Brennan's monorail car of 1916 there is applied about the precession axes of the gyros a torque which, at any instant, is nearly proportional to the velocity of precession, to the angular displacement of the spin-axles from the central position, and to the frictional torque opposing precession.

Figure 219 shows the two gyros, as seen by an observer looking down upon them, spinning in opposite directions about axes normally horizontal and perpendicular to the track, and capable of precessing about vertical axes. Rigidly attached to the gyro-casings G and G' are gear segments g and g'. These gear segments mesh with a double rack r connecting two pistons capable of being pushed back and forth by air pressure in two cylinders C and C'.

* Eddy, "The Mechanical Principles of Brennan's Monorail Car," J. Franklin Inst. (1910), p. 467.

The admission of compressed air into these cylinders and the exhaust from them are controlled by a compound air valve, not shown in the diagram, operated by a device attached to the upper end of the precession axle D of one of the gyros.*

As soon as the car tilts, the spin-axles of the two gyros begin to precess in opposite directions. The rotation of the precession axle of G' causes the attached mechanical device to exert a force on the end of a rod, to the other end of which is a compound air valve connected into the service lines of the servomotor cylinders. This force is the resultant of a force that is proportional to the precessional velocity, to a force that is proportional to the angular displacement of the gyros from their central position, and to the friction around the precessional axes. This resultant force opens the valves of the servomotor to a degree proportional to the force, thereby accelerating the precessional velocity by an amount proportional to the torque that tilted the car from the vertical. This acceleration of the precessional velocity is accompanied by a torque adequate to erect the car if the dimensions of the apparatus are of the proper values. The car is furnished with a device to hold the car upright when the gyros are not spinning.†

164. The Scherl Duogyro Monorail Car of 1912. — The stabilizing device consists of two gyros, spinning in opposite directions about axes that normally are vertical, and capable of precessing about axes that are transverse to the longitudinal axis of the car. The two gyro-casings are coupled together by a pair of gear segments so that the velocities of precession of the two gyro-axles, at any instant, are equal in magnitude and opposite in direction. There is a motor that accelerates the precession at the proper times. This precession motor is of the hydraulic type, consisting of a cylinder and piston together with a set of valves that are controlled by the rocking movements of the gyro system. Hydraulic pressure is produced by an electrically driven pump. A device‡ is provided so that, in case there should be an interruption of the main current that operates the gyro and precession motors, the continued rotation of the gyros will cause the gyro-motors to act as electric generators, the current thereby produced operating the stabilizing system for a considerable time. As soon as the speed of rotation of the gyros falls to such a value that the stability

---

* U. S. Patent. Brennan, No. 1183530, 1916.
† U. S. Patent. Brennan, No. 1019942, 1912.
‡ U. S. Patent. Falcke, No. 1048817, 1912.

of the car is endangered, a magnetic device releases a set of supports that will hold the car upright.*

When the car tilts to one side, the gyro spin-axles precess toward one another; when the car tilts to the other side the spin-axles precess away from one another. In either case, the tilting of the spin-axles operates valves which control the operation of the hydraulic precession motor piston. The movement of this piston accelerates the precessional velocity of the two connected gyros. The result is a torque which tilts the car upward and, possibly, past the vertical position.

* U. S. Patent. Scherl, No. 959077, 1910.

INDEX

(The numbers refer to pages)

A

Acceleration, angular, 4, 10
  linear, 1, 7
  radial, 2
  relation between angular and linear, 11
Agonic line, 165
Airplane, gyroscopic actions on, 51, 59
Airplane cartography, 100
Airplane gyroscopic pilot, 123
Angular momenta, 5, 10, 17, 20
Anschütz gyro-compass, 224–235
Anschütz gyro-horizon, 117
Anti-roll devices for ships, 42, 131–161
Arma gyro-compass, 236–247
Autogiro, 52

B

Ballistic damping or turning error, 190
Ballistic damping or turning eliminator, 212, 246
Ballistic deflection error, 184, 186
  avoided, 210, 222, 231, 244, 252
Beats, 38
Bessemer's steamer with gyroscope, 62
Bliss-Leavitt torpedo stearing gear, 97
Bonneau airplane inclinometer, 76
Bonneau-LePrieur-Derrien gyro-sextant, 119
Brennan monorail cars of 1905 and 1916, 268
Brown gyro-compass, 170
Bumstead sun compass, 170

C

Camera control on an airplane, 101–104
Cardan suspension, 45
Carter's track recorder, 125
Cartography, airplane, 100
Center of gravity, 21
Center of mass, 21
Centrifugal drier, Weston's, 75
Centripetal and centrifugal forces, 12, 16
Centroid, 21
Clinging of spinning body to guide, 85
Clinograph for measuring crookedness of well casings, 74
Compass, gyroscopic, 162
  inductor, 169
  magnetic, 165
  magneto, 169
  sun, 170
Couple, force, 4
Coupled systems, 36
Course and speed errors of gyro-compasses, 182
  avoided, in Anschütz compass, 231
  in Arma compass, 241
  in Brown compass, 220
  in Florentia compass, 251
  in Sperry compass, 206

D

Damping of vibrations, 39
  in Anschütz gyro-compass, 228
  in Arma gyro-compass, 237
  in Brown gyro-compass, 218
  in Florentia gyro-compass, 250
  in Sperry gyro-compass, 203
Degrees of freedom, 44

274 INDEX

Deviations of magnetic compass, 166, 168
Directed gun-fire control, 128
Directorscopes, gun-fire, 129
Drift of projectiles, 80
Dynamic stability, defined, 255
  some laws of, 257
Dyne, 2

E

Earth inductor compass, 169
Earth's axis and spin-axis of gyroscope, 71–74
Energy, 2
  of precessing body, 66
Erg, 2
Errors to which a gyro-compass is subject, 182–196

F

Fieux ship stabilizer, 144
Fleuriais gyroscopic octant, 120
Florentia gyro-compass, 247–253
Follow-up repeater system of Anschütz gyro-compass, 234
  of Arma gyro-compass, 238
  of Brown gyro-compass, 221
  of Florentia gyro-compass, 248
  of Sperry gyro-compass, 202
Force, 2, 5
  centripetal and centrifugal, 12, 16
Frahm anti-roll tanks, 42
Freedom, degrees of, 44
Friction at peg of top causes rise of spin-axis, 76

G

Gimbals, 45
Griffin pulverizing mill, 92
Gun-fire control, 128
Gun-fire control compasses, 196
Gun-fire control directorscope, 129
Gyration, radius of, 14
Gyro, gyroscope, gyrostat, defined, 45
Gyro-compass, Anschütz, 224–235
  Arma, 236–247
  Brown, 215–223

Gyro-compass, Florentia, 247–253
  meridian-seeking tendency of, 173–181
  Sperry, 198–214
Gyro-compass is subject to errors, 182–196
Gyrodynamics, first and second laws of, 55
Gyro-horizontals, 115–128
Gyro-pendulum, 105
  of the same period as a certain simple pendulum, 113
Gyroscope modifies rolling of a ship, 140
Gyroscopic conical pendulum, 105–108
Gyroscopic torque or resistance, 49
Gyro-verticals, 115

H

Horizon, Sperry airplane, 78
Horizontals, dynamic and true, 12, 13
Horse-power, 3
Hurrying and retarding precession, 69

I

Inclinometer, Bonneau's, 76
Inductor compass, 169
Inertia, 3
  moment of, 4, 13–15
Intercardinal error, of gyro-compasses, 191
  suppression of, 195, 212, 222, 232, 247, 252

K

Kinetic energy of a precessing body, 66
Kinetic stability, defined, 255
  of monorail cars, 259–272
  some laws of, 257

L

Latitude, determination of, 115
Latitude error of gyro-compasses, 182, 205

INDEX

Latitude error of gyro-compasses,
    avoided, 220, 231, 241
    corrected, 206, 251
Leavitt torpedo steering gear, 97
Locomotive, gyroscope couple acting on, 63–66

M

Magnetic compass, deviations, produced by a rapid turn, 168
    when on an iron ship, 166
    directive tendency of, 165
Magneto compass, 169
Mass, 3
Maxwell top, 10
Meridian, magnetic, 162
Meridian method of locating the geographic, 162
Meridian-seeking tendency, of a liquid-controlled gyroscope, 176
    of a magnetic compass, 165
    of a pendulous gyroscope, 173
Meridian-seeking torque acting on a gyro-compass, 178
Meridian-steaming error, of gyro-compasses, 182
    Anschütz, 231
    Arma, 241
    Brown, 220
    Florentia, 251
    Sperry, 206
Metacentric height, 132
Moments of inertia, 4, 13, 15
    dynamic, 14
    principal, 15
Momentum, angular, 5, 10, 17, 20
Monorail cars, gyroscopic stabilization of, 257–272
    Brennan, 268
    Scherl, 271
    Schilovsky, 266
Motorcycle, gyroscopic forces acting on a, 57

N

North-steaming error, of gyro-compasses, 182
    Anschütz, 231

North-steaming error, Arma, 241
    Brown, 220
    Florentia, 251
    Sperry, 206
Nutation, 67

O

Octant, Fleuriais gyroscopic, 120

P

Pendulous gyroscope, 105
    meridian-seeking tendency of, 173
    period of, 108–111
Pendulous mass, 191
Pendulum, conical, 29
    forces acting on an unsymmetrical, 16
    mathematical or simple, 29
    physical, 28
    with attached gyroscope, 136, 146
Period, of a conical pendulum, 30
    of a gyro-compass, 186–190
    of a mathematical and of a physical pendulum, 29
    of simple harmonic motion of rotation, 25
    of the roll of a ship, 134
Phase and phase angle, 32
Pioneer turn indicator, 85
Pitching of a ship due to waves, 132
Port side of a ship, defined, 50
Power, 2
Precessing body, kinetic energy of, 66
Precession, 48
    period of, 66
Precessional torque, 49
Precessional velocity maintained by a torque, 55
Projectile, drift of a, 80

Q

Quadrantal deflection, 191, 195
Quadrantal error, avoided, in the Anschütz gyro-compass, 232
    in the Arma gyro-compass, 247
    in the Brown gyro-compass, 222

Quadrantal error, avoided, in the Florentia gyro-compass, 252
in the Sperry gyro-compass, 212

R

Radian, 3
Repeater system, of the Anschütz gyro-compass, 234
of the Arma gyro-compass, 238
of the Brown gyro-compass, 221
of the Florentia gyro-compass, 248
of the Sperry gyro-compass, 213
Resonance, 36
Retarding and hurrying of precession, 69
Revolving a gyroscope having two degrees of freedom, 82
Roll and pitch recorder, 121
Rolling error, 191
avoided, in the Anschütz gyro-compass, 232
in the Arma gyro-compass, 247
in the Brown gyro-compass, 222
in the Florentia gyro-compass, 252
in the Sperry gyro-compass, 212
Rolling of a ship, by waves, 131
methods of diminishing, 135
period of, 134
produced by a gyroscope, 152
Rotation, simple harmonic motion of, 25

S

Scherl's monorail car of 1912, 271
Schlick's ship stabilizer, 142
Schilovsky's monorail car of 1915, 266
Schuler's gyro-horizon, 117
Sextants, gyroscopic, 119, 121
Ship stabilizer, of the active type, 146
Sperry's, 146–160
of the inactive type, 136
Fieux, 144
Schlick, 142
Side-wheel steamer, gyroscopic torque acting upon, 50

Simple harmonic functions, mean value of products of, 33
Simple harmonic motion, of rotation, 25
of translation, 22
Slug, 3
Sperry-Carter track recorder, 125
Sperry's airplane horizon, 78
Sperry's airplane pilot, 123
Sperry's gyro-compass, 198–215
Sperry's roll and pitch recorder, 121
Sperry's ship stabilizer, 147–160
Spin-axis of a gyroscope and the earth's axis, 71, 74, 172
Spinning, defined, 44
Stability, static and kinetic, 255
Stabilization by gyroscopes, 255–272
Starboard of a ship, defined, 50
Steamship, gyroscopic torque acting on, 63
Sun compass, 170
Swing radius, 14

T

Tanks, Frahm anti-roll, 42
Taylor's formula for ship stabilization, 153
Torpedo control, 94–100
Torque, 4, 10
Torque and angular momentum, 17
Torque developed by rotation of the spin-axis, 60
Torque effect on a spinning body, 45, 52, 87, 90
Torque opposing turning of the second frame of a gyroscope, 111
Torque required to maintain constant precession, 52
Top, rising of the spin-axis of, 76
Track recorder, Sperry-Carter, 125
Turn indicator, Pioneer, 85

V

Vertical, the true and the dynamic, 12, 13
Velocity, angular, 4, 10

Velocity, linear, 1, 7
  relation between angular and linear, 11
Vibrations, damping of, 39
  free and forced, 36

W

Wave motion, 30, 131–135
Weston centrifugal drier, 75